Defend and Befriend

Defend and Befriend

The U.S. Marine Corps
and
Combined Action Platoons
in Vietnam

John Southard

UNIVERSITY PRESS OF KENTUCKY

Scholarly publisher for the Commonwealth,
serving Bellarmine University, Berea College, Centre College of Kentucky, Eastern
Kentucky University, The Filson Historical Society, Georgetown College, Kentucky
Historical Society, Kentucky State University, Morehead State University, Murray
State University, Northern Kentucky University, Transylvania University, University of
Kentucky, University of Louisville, and Western Kentucky University.
All rights reserved.

Editorial and Sales Offices: The University Press of Kentucky
663 South Limestone Street, Lexington, Kentucky 40508-4008
www.kentuckypress.com

Maps by Richard A. Gilbreath, University of Kentucky Cartography Lab.

Library of Congress Cataloging-in-Publication Data

Southard, John, 1980–
   Defend and befriend : the U.S. Marine Corps and Combined Action Platoons in Vietnam /
John Southard.
      pages cm
   Includes bibliographical references and index.
   ISBN 978-0-8131-4526-6 (hardcover : alk. paper) — ISBN 978-0-8131-4527-3 (pdf) —
ISBN 978-0-8131-4528-0 (epub)
   1. Vietnam War, 1961–1975—Civilian relief. 2. United States. Marine Corps—History—
Vietnam War, 1961–1975. 3. United States. Marine Corps—Civic action. 4. Combined
operations (Military science) I. Title. II. Title: U.S. Marine Corps and Combined Action
Platoons in Vietnam.
   DS559.63.S68 2014
   959.704'345—dc23   2014005216

To my wife, Rachel,
and my parents, Dan and Kari

Contents

Illustrations follow page 104

SOUTH VIETNAM

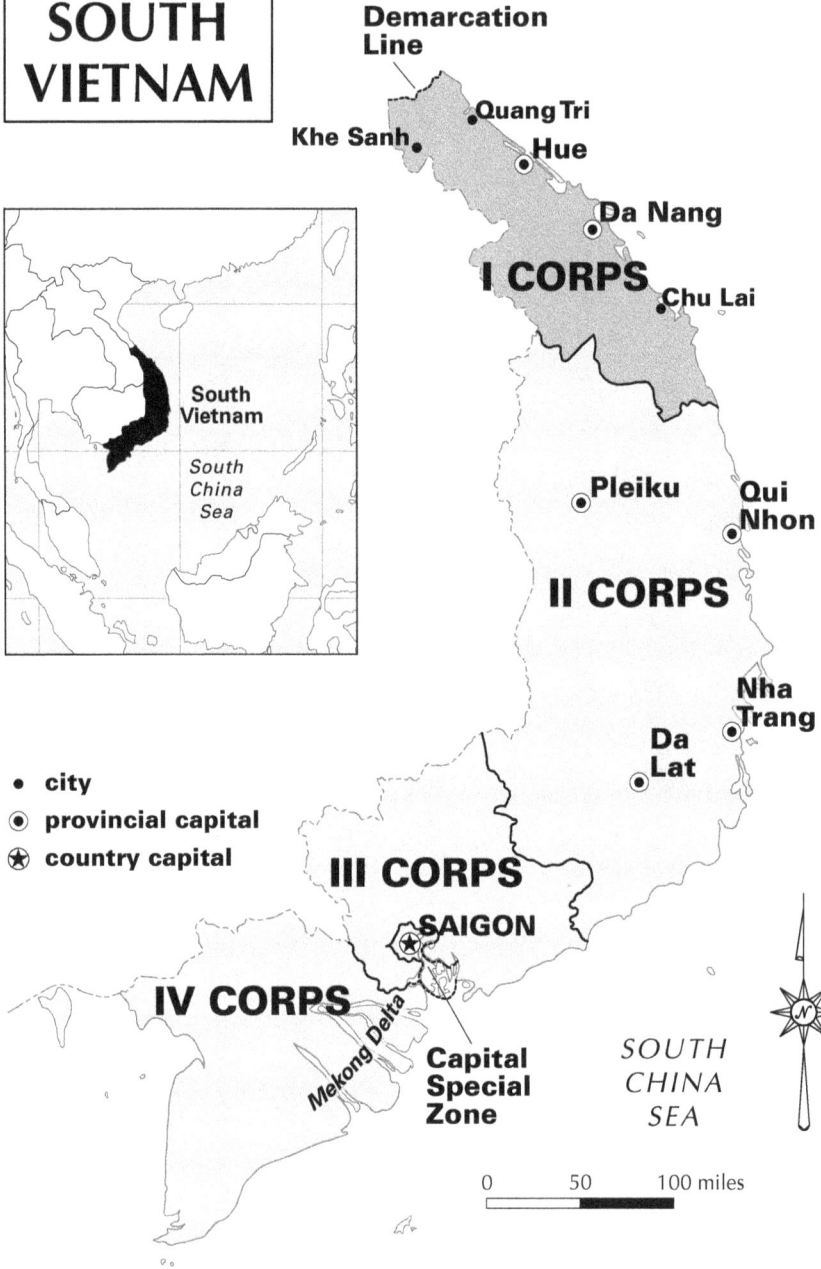

Demarcation Line

Quang Tri

Khe Sanh

Hue

Da Nang

I CORPS

Chu Lai

South Vietnam

South China Sea

Pleiku

Qui Nhon

II CORPS

Nha Trang

Da Lat

- • city
- ◉ provincial capital
- ✪ country capital

III CORPS

SAIGON

IV CORPS

Mekong Delta

Capital Special Zone

SOUTH CHINA SEA

0   50   100 miles

Corps military regions in South Vietnam, 1970.

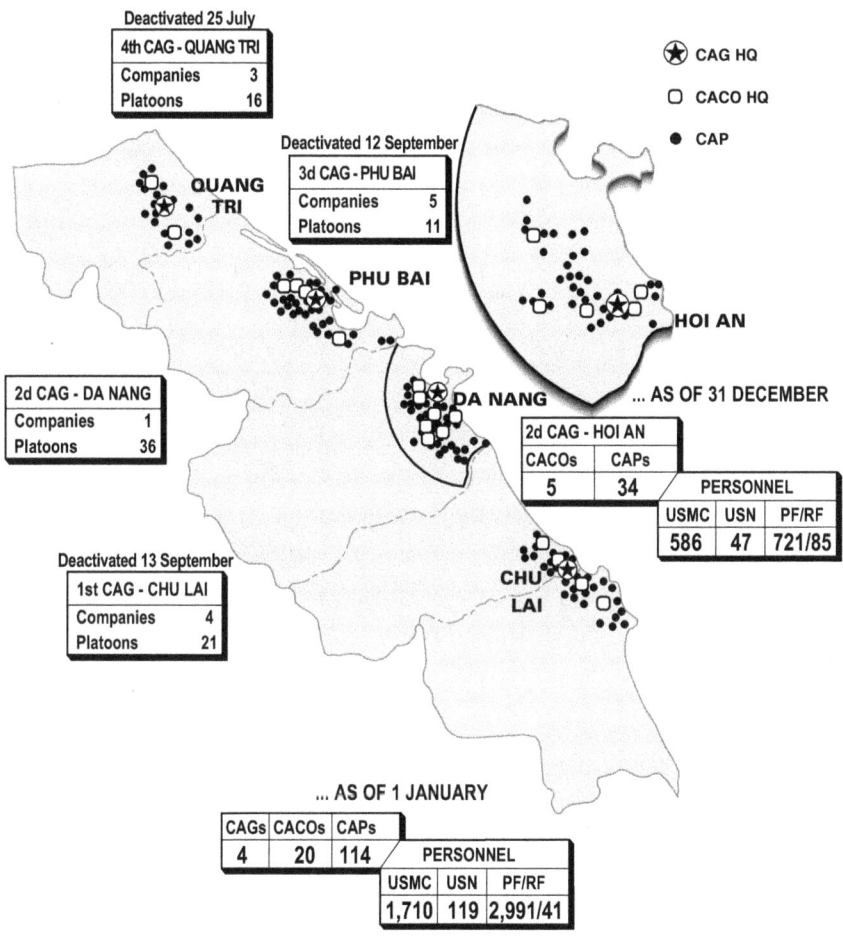

Evolution of the Combined Action Program, 1970.

Preface

My interests in the history of the U.S. Marine Corps and the Vietnam War stem from my dad, a former Marine and Vietnam veteran. Although my dad has remained mostly silent about his wartime experiences, he has never shied from declaring his unwavering pride in the Marine Corps. When I was in grade school, sleeping in on Saturday mornings was seldom an option for my sisters and me, as my dad shouted his own Marine Corps boot camp version of "Reveille." Apparently, eight o'clock was too late to sleep on a Saturday. Socially reserved and humble, my dad is also brutally honest. An alumnus of Texas A&M University once gloated to my dad about the "Corps of Cadets," the individuals who comprise that institution's ROTC program. With a stone-faced glare my dad responded, "There is only one Corps." The adage "Once a Marine, always a Marine" aptly applies to my father.

There is a distinct culture deeply engrained in the U.S. Marine Corps and the individuals who hold the title of U.S. Marine. The smallest of all U.S. military branches, the Marine Corps has embraced its image as a rough and tough, loud and proud group of supremely confident riflemen who stand ready to annihilate America's enemies anytime and anywhere. In his *Making the Corps,* a masterful look into the culture of the Marine Corps, Thomas Ricks shares an "old Pentagon" joke about the different U.S. military services: "Each service is told to 'secure' a building. The Marine Corps wants to destroy it, the Army wants to establish a defensive perimeter, the Navy wants to paint it, and the Air Force wants to lease it for five years."[1] As an adolescent and young adult, I always associated annihilation and destruction with the U.S. Marine

Corps and its band of "leathernecks" and "devil dogs." As an under-graduate, I wore a U.S. Marine Corps T-shirt with the words "When it absolutely, positively has to be destroyed overnight" blazoned across the chest. For much of my life, these images and beliefs stoked my perception of the Marines.

When I arrived at Texas Tech University in 2006, the Vietnam War quickly ascended to the top of my list of general topics for a book project. At Texas Tech, I had immediate access to the Vietnam Center and Archive, the largest archive on the Vietnam War outside of the National Archives in Washington, more than a thousand miles from Lubbock, Texas. However, browsing through the millions of pages in the Vietnam Archive still did not yield a specific topic. Moreover, scholars had already meticulously detailed seemingly every military-related aspect of American involvement in Vietnam. After my first academic year at Texas Tech, I had no idea what specific avenue I would take for my research.

In the summer of 2007, I traveled through Vietnam as part of the Vietnam Center's study abroad program. The trip was a life-changing experience. During our three-week tour, we ventured from Hanoi to the country's southernmost reaches in the Mekong Delta, making numerous stops along the way. I stood on historically significant grounds that had played critical roles in the war. Images from places such as Ap Bac, Khe Sanh, the beaches of Da Nang, and the former presidential palace of South Vietnam will forever be etched into my memory, as will the phone conversation I had with my father from the area he had patrolled during the war. As I gazed across the beautiful and peaceful landscapes of Vietnam, it was hard to imagine that violence, death, and devastation had once permeated these areas.

In addition to acclimating myself to the geography, topography, and climate of Vietnam, I also gained a better understanding of the Vietnamese people. Among the dozens of urban and rural areas to which we ventured, the villages offered the most penetrating glimpse into traditional Vietnamese culture and society. The villages were void of the amenities that Americans often take for granted—refrigeration, toilets, air-conditioning, and clean drinking water. Yet I was awestruck by the positive and gracious attitudes of the villagers. Immersed in an environment that most Americans would deem abysmal and otherworldly, the villagers were calm, patient, and most of all happy. I realized firsthand that regardless of adverse circumstances, the villagers simply wanted to live peacefully in their ancestral homes with their families.

After my study abroad trip, I ultimately chose the Marine Corps' combined action platoons (CAPs) as my book topic. The concept of Marines trained to annihilate and destroy living side by side with Vietnamese villagers intrigued me, jostling my preconceived perceptions of leathernecks and devil dogs. Assignment in a combined action platoon required a combination of patience, sensitivity, and empathy, attributes seldom linked with U.S. Marines. As I dug deeper into the Combined Action Program, I was stirred by the thought of teenaged Marines living in Vietnamese villages, interacting with the people. I presumed that my future research would show that the Marines and corpsmen could not overcome the seemingly impenetrable barriers, most notably the cultural ones, that existed between them and the villagers. However, as I dug deeper, I found I was wrong. Many of the Marines and corpsmen actually grew fond of their villages and the people. Images of CAP Marines playing baseball and volleyball with Vietnamese village children now blurred my perception of the Marine Corps as an institution dedicated solely to annihilating.

While I researched and wrote this book, I often thought about the Vietnamese villagers with whom I spoke. During the Vietnam War, competing armed forces constantly jockeyed for control of the villages and the people who inhabited them. Villagers were uprooted from their homes and forced into military service away from their families. Armies ransacked their villages, burned their homes, slaughtered their animals, and murdered their family members. This is why the study of combined action platoons fascinates me. The U.S. Marine Corps devised a counterinsurgency program that not only protected villagers but allowed them to stay near their families, an idea that stood in stark contrast to South Vietnamese village relocation programs and American "search and destroy" missions. Unlike many South Vietnamese– and American-orchestrated pacification programs, combined action platoons featured Americans actively engaging with the daily lives of villagers, sincerely attempting to respect their customs. CAP Marines and corpsmen were the outsiders. Rather than forcing villagers to adhere to American military demands, Americans in combined action platoons had to adjust to the Vietnamese way of life.

CAP veterans meet annually for a reunion, the dates always encompassing 10 November, the birthday of the Marine Corps. On that evening, about half the veterans attend the formal Marine Corps ball, donning either Marine "dress blues" or a tuxedo, with the "fruit salad" of their earned decorations and awards pinned to the left breast of their outerwear. Among the numerous events that take place in the designated "hospitality room" of the reunion site, the CAP veterans reserve one night to commemorate the birthday of the Marine Corps. In adherence to Marine Corps tradition, the oldest and youngest Marines in the room have the honor of cutting the cake, which is followed by a loud and proud rendition of "The Marines' Hymn." The veterans are proudest of all that they were CAP Marines.

In 2010 and 2011, I attended reunions of combined action platoon veterans in Washington, DC, and San Antonio, Texas, respectively. The reunion committee graciously approved my request to conduct interviews. The CAP veterans at both reunions were thrilled that someone outside their group had an interest in the program. Sprinkled among CAP Marine veterans were the corpsmen, the vital links that kept them alive. Each CAP corpsman served as the primary medic for both the Americans and the villagers. Corpsmen treated ailments and wounds, procured and distributed medical supplies, and accompanied Marines and Popular Forces (PF) soldiers on patrols. In the hospitality room, the veterans congregated on a nightly basis, sometimes calling for "Doc" when they vied for the attention of one of the former corpsmen. Two topics seemed to dominate their conversations: patrolling outside the village perimeter at night with only five or six other Americans, and anecdotes about the villagers, especially the children, whom many of the veterans befriended. Veteran CAP Marines and corpsmen alike know that what they did in the war sets them apart from the regular land-based units in South Vietnam.

Abbreviations and Acronyms

AIT	advanced infantry training
ARVN	Army of the Republic of Vietnam
CAC	combined action company; also known as CACO
CACO	combined action company; also known as CAC
CAG	combined action group
CAP	combined action platoon
CARE	Cooperative for American Relief Everywhere
CIA	Central Intelligence Agency
CIDG	Civilian Irregular Defense Group
CINCPAC	commander in chief, Pacific
COMUSMACV	commander, U.S. Military Assistance Command, Vietnam
CORDS	Civil Operations, Revolutionary Development Support
CRS	Catholic Relief Services
CUPP	Combined Unit Pacification Program
DMZ	demilitarized zone
DRV	Democratic Republic of Vietnam
FM	field manual
FMFPAC	Fleet Marine Force, Pacific
FULRO	United Struggle Front for the Oppressed Races
FWMF	Free World Military Force
GVN	Government of Vietnam
HES	hamlet evaluation system
ICIPP	Infantry Company Intensified Pacification Program

IED	improvised explosive device
JCC	Joint Coordinating Council
JCS	joint chiefs of staff
JUSPAO	Joint U.S. Public Affairs Office
KIA	killed in action
MAAG	Military Assistance and Advisory Group
MACCORDS	Military Assistance Command, Vietnam, Civil Operations, Revolutionary Development Support
MACV	Military Assistance Command, Vietnam
MAF	Marine Amphibious Force
MAT	mobile advisory team
MEDCAP	medical civic action patrol
MEDEVAC	medical evacuation
MOS	military occupational specialty
MTT	mobile training team
NCO	noncommissioned officer
NGO	nongovernment organization
NLF	National Liberation Front
NVA	North Vietnamese Army; also known as People's Army of Vietnam (PAVN)
OCO	Office of Civil Operations
PAVN	People's Army of Vietnam; also known as North Vietnamese Army (NVA)
PF	Popular Forces
PLAF	People's Liberation Armed Forces
PROVN	*Program for the Pacification and Long-Term Development of South Vietnam*
PSDF	People's Self-Defense Forces
RD	revolutionary development
RF	Regional Forces
RVN	Republic of Vietnam
RVNAF	Republic of Vietnam Armed Forces
SEATO	Southeast Asia Treaty Organization
SF	Special Forces
TFES	territorial forces evaluation system
TOE	table of organization and equipment
USAID	U.S. Agency for International Development
USOM	U.S. Operations Mission
VC	Viet Cong

Introduction

Up to that point they were gooks. After that they were Vietnamese.
—Tom Morton

Thus Tom Morton, a former U.S. Marine corporal in the Vietnam War, succinctly expressed how his perception of the Vietnamese people changed during his tenure in a combined action platoon (CAP). Morton, a squad of his fellow Marines, and a U.S. Navy corpsman lived in a South Vietnamese village for months, training the local militia, conducting twenty-four-hour patrols, and rendering civil and medical aid to civilians, all to keep the area free of enemy influence and control. Morton's statement may come as a surprise to students of the Vietnam War who have read some of the countless published memoirs and oral histories of American veterans. Those in the general public who have spoken with Vietnam veterans about their experiences would also likely find Morton's assertion about the Vietnamese uncommon. Some of the most popular firsthand accounts of the war reveal that the American military viewed the Vietnamese through a racist lens. Many Americans' perceptions of the Vietnamese as subhuman never wavered during their tours in Vietnam. Fighting a war of attrition against guerrillas proved to be a frustrating experience for U.S. infantrymen, or "grunts," in Vietnam. Told to search for and destroy an elusive enemy hidden in the jungle or disguised as civilians in the villages, American infantrymen frequently patrolled for days without seeing one guerrilla combatant. As the days without enemy contact mounted, many American infantry units unleashed their frustrations with deadly force against villages and their civilian occupants. The irritated grunts often assumed the civilians were the enemy they had failed to find on patrol.

Many Americans who joined the Combined Action Program, as Morton's quote suggests, arrived in their assigned villages with a racist perception of the Vietnamese people. Starting in boot camp, drill instructors dehumanized the enemy that American recruits would encounter in Vietnam. Racial slurs such as "gook" and "slant eyes" characterized the discourse of American military personnel before, during, and after the war. Yet these CAP Marines and corpsmen lived in South Vietnamese villages, leaving the hamlets when the local militia, the Popular Forces (PF), had proven they could operate effectively without the Marines' assistance. The small group of Americans in each CAP village had to adapt to the Vietnamese way of life. Personal and unit survival depended heavily on creating and maintaining an amicable rapport with the villagers, for the Vietnamese civilians held information about recent and upcoming enemy movements and attacks near the villages. Moreover, the traditional Vietnamese village housed generations of families living in close quarters, which meant that many of the civilians were cognizant of who in the cluster of hamlets supported the Viet Cong (VC) or had outright joined it.[1] The most effective and humane means of extracting intelligence from the Vietnamese was for the Americans to make the indigenous peoples comfortable with their presence, which proved a daunting task for the Marines when they initially entered the villages. In other words, the Marines and corpsmen had to defend and befriend people whom many had already learned to hate. Yet by the time of the Marines' departure from their assigned villages, many of the Americans had accepted dinner invitations to village elders' homes, attended festivals and weddings, and taken part in funeral processions for PF and Americans in the CAPs who had recently fallen in battle. This book seeks to explain how and why the program as a whole and the individual Marines and corpsmen in the villages attempted to overcome military and cultural obstacles in CAP villages.

As part of the U.S. military's efforts to win the "hearts and minds" of the people, the CAP Marines performed civic action duties to bolster village infrastructures, most often consisting of the distribution of sanitary supplies and helping to procure materials to build pig pens, wells, office buildings, schoolhouses, and the like. In the process of implementing civic action, the Marines began to interact with the Vietnamese. However, more than any other element of a CAP, the daily medical services provided by the corpsmen, known as medical civic action patrols (MED-CAPs), helped to create and maintain cordiality between the Americans and the villagers. At a central location in the villages, the corpsmen remedied common ailments such as headaches and stomach viruses, and in some cases treated life-threatening injuries. At times, villagers lined up by the hundreds to receive treatment from the corpsman during his routine MEDCAP. Although the MEDCAPs were the staple of the corpsman's medical responsibility for the civilians, he also tended to the Marines' and villagers' medical needs twenty-four hours a day. Indeed, the corpsmen played a critical role in igniting the cross-cultural exchanges that ultimately transpired in CAP villages, making them perhaps the most indispensable Americans in CAPs.

"New" military history has emphasized the experiences of military personnel on the front lines but has not stressed the inevitable encounters between the occupying belligerents and the civilians of the occupied territory. This book reconceptualizes the American military experience in Vietnam, allowing us to reexamine common perceptions of infantrymen in the conflict. We are used to seeing in secondary sources and memoirs the ways that many American military personnel disrespected and demonized the Vietnamese people and their culture as a result of their experiences in war. My work presents a different perspective: how thousands of Americans tasked with counterinsurgency duties came to perceive the Vietnamese in a more positive light after months and in some cases years of daily interaction.

Due to the difficulty in obtaining Vietnamese sources specifically pertaining to the Combined Action Program, this book is necessarily an analysis of CAPs from the American perspective. Only one publication provides the Vietnamese perspective of life in a CAP village. James Trullinger's *Village at War* examines the evolution of the village of My Thuy Phuong from the French colonial period through U.S. military intervention.[2] From 1969 to 1972, Trullinger worked for the U.S. Agency for International Development (USAID) in Saigon and Da Nang. After a brief stint in the United States, Trullinger returned to South Vietnam in 1974 to conduct a study of the effects of decades of war on My Thuy Phuong. From November 1974 to March 1975, just before the Communist takeover of Saigon in late April, Trullinger conducted 175 interviews with the villagers about their experiences over the previous four decades. During the U.S. military presence, the Marines dispatched a CAP to My Thuy Phuong, but the wide chronological period covered in Trullinger's book does not allow for a lengthy description of the villagers' reactions to the Americans' sustained presence. Throughout my work, descriptions of the villagers' reactions to the Americans' presence derive from the Marines and corpsmen.

To understand the nature of the entire program from top to bottom, one must acknowledge the perspectives of the colonels and generals who managed and organized CAPs as well as the Marines and corpsmen who lived in the villages. My research has shown that reports and comments from the colonels and generals often did not match the experiences of the enlisted men and NCOs in CAP villages. Ron Milam's *Not a Gentleman's War* debunks the myth held by many colonels that the American junior officer corps in Vietnam was plagued with soldiers who lacked commitment and competence.[3] Milam shows that the conduct of U.S. Army lieutenant William Calley, the junior officer in command during the My Lai Massacre in 1968, was the exception, not the rule, in the war. In the appendix of his book, Milam reveals that in the immediate years succeeding the fall of Saigon in 1975, numerous "angry colonels" in the U.S. military wrongly blamed the junior officer corps for losing the war. As Milam has articulated, colonels and generals have perceived and remembered the war much differently than the troops in the field who followed the orders, pulled the triggers, and saw the enemy face-to-face. One of the most glaring discrepancies within the program comes from descriptions of PF in the villages, detailed in chapter 5. While top-ranking Marines constantly gave optimistic reports of the PF's progression in CAPs, testimony from the Marines who conducted the hands-on training reveals a local indigenous force that sluggishly and reluctantly performed its duties.

General military histories of the Vietnam War often neglect or at times outright ignore the experiences of the men and women in the bottom rungs of the U.S. military's hierarchical rank structure. With the recent onslaught of "new" military history, many scholars have skillfully reversed the tendency to focus solely on high-ranking commanders. However, it is important that military history continue to analyze the colonels and generals, for they help formulate the strategy that in part dictates how their subordinates conduct the war. This especially applies to the Vietnam

War, where American generals initiated an annihilation strategy against an enemy that chose a different strategic approach. Ordered to search for enemy main force units and destroy them on sight, American infantry-men often wandered through the jungles of South Vietnam for days with-out making contact, while their adversaries patiently and intentionally eluded U.S. forces until they possessed a clear advantage.

With the exception of the first combined unit in August 1965, there were no American officers in the villages aside from the occasional visit from an officer based at the district headquarters, miles away from the CAPs. Only enlisted Marines and NCOs lived in the villages. The senior ranking NCO generally served as the commander of the Marines, but in the absence of the sergeant, one of the enlisted Marines assumed control of the unit. In July 1969, with the program just one month and three units shy of peaking at 114 CAPs, the average CAP leader was twenty-one years old. That same month, nearly 40 percent of all CAP leaders carried the enlisted rank of corporal or lance corporal, which meant that many of the combined units had either eighteen- or nineteen-year-old Marines serving as commanders.[4] Rarely in military history does one find a situa-tion in which teenaged enlisted Marines are given command of a unit that demanded such patience, maturity, and cultural adaptability as duty in a CAP did. According to Lt. Col. William Corson, the director of the pro-gram in 1967, officers were absent from CAPs because "the people associ-ate them with corruption. But our young corporals and sergeants look like they have succeeded in the same way that the best of the village boys, who can only aspire to becoming corporals and sergeants, might hope to suc-ceed."[5] Corson, credited with bolstering the efficiency and organization of the program, believed that implanting nineteen- and twenty-year-old Marines in the villages would net positive results because many of them lacked a high school education and had "nothing to look forward to."[6]

Corson's desire to select Marines with at most a high school education was not unique in the Vietnam-era military. Eighty percent of the men who ventured to Vietnam had no education beyond high school, and the Marine Corps corralled many recruits from this demographic.[7] After join-ing the Marines in 1967, James Stanton recalls that "the only people that would take me was the Marine Corps because I never had a high school education."[8]

Perhaps the reasons for targeting this demographic expand beyond the belief that these individuals "had nothing to look forward to." Kyle Longley's *Grunts: The American Combat Soldier in Vietnam* offers a perceptive analysis of how social and cultural constructs in America in the 1950s and 1960s helped to shape the perceptions of U.S. combat forces in Vietnam. The baby boomers who would fight in Vietnam grew up in a cultural environment that applauded the masculinity of military service embodied by the veterans of the Second World War. The veterans of World War II and Korea, who were fathers, uncles, teachers, and coaches in the late 1950s and early 1960s, served as role models for the adolescent males who would serve in Vietnam. Longley contends that many of the men who saw combat in Vietnam came from blue-collar neighborhoods and agricultural communities where physical skills rather than intellectual prowess defined masculinity. The Marine Corps seemed a perfect fit for those seeking to prove and advance their masculine stature. Among all of the U.S. military branches, Longley argues, the Marine Corps "ranked at the highest in terms of toughness."[9]

The vast majority of CAP Marines held an infantry military occupational specialty (MOS), but in times of need, the program gladly accepted troops from noninfantry units without combat experience.[10] From 1965 to 1969, the program accepted volunteers from mostly infantry units in the field. Yet with the mounting ferocity of the war in the northernmost provinces of South Vietnam, mainline Marine units were needed as close to full strength as possible, which left the program searching for more bodies to place in CAP villages. In 1969, the program appointed Marines stationed in the United States to CAP duty before their arrivals in South Vietnam. None of those Marines had prior combat experience. Whether Americans entered the program from the United States or came from infantry or noninfantry outfits in Vietnam, all of them needed an introduction to the distinct military and cultural environments in a CAP village. For those joining the program in South Vietnam, nothing they had experienced prior to their CAP assignment had prepared them for the adjustments they needed to make in the villages.

Beginning in 1967, the program formalized a school in Da Nang where all Marines and corpsmen received instruction on the military and cultural intricacies of CAP duty before landing in the villages. For two weeks, the CAP recruits learned small-unit military tactics in addition to taking an introductory course on Vietnamese language and culture.

Linguistic instruction focused on military-related words and phrases that would allow the Americans to communicate with PF when either planning or executing a patrol. The language and culture instruction proved to be the most critical lessons of CAP school. However, in just two short weeks, the Americans were unable to gain proficiency in either. For the Americans in the villages, living among the Vietnamese and interacting with them furnished the best language and culture training.

Americans entering a newly created CAP faced an awkward and uncomfortable social environment. In time, however, through civic action and daily medical calls from the corpsmen, many villagers began to open up to the Americans. As interaction increased between the Americans and Vietnamese in CAPs, the villagers gradually began to divulge intelligence to the Marines.

Militarily, the Marines had to train the PF, the most marginalized element of the South Vietnamese military. Members of the PF, most of whom operated in their home villages, generally received lackluster military training, and throughout the program's tenure they exhibited low morale and a reluctance to accompany Marines on patrols. Although statistics reveal an increase in the effectiveness of CAP PF as the war progressed, the Marines who worked with the local forces experienced considerable frustration and in some cases suspected them of supporting the VC. While some CAP Marines remember working with competent individuals determined to thwart VC efforts to control their villages, others recall incompetence, indifference, and ineffectiveness from the PF.

American and South Vietnamese civilian and military leaders hoped the PF, who theoretically represented the Republic of Vietnam (RVN), could connect the villagers politically with the South Vietnamese government in Saigon. With nearly three thousand villages spread across South Vietnam, the RVN largely relied on the PF operating in their home villages to spread support for its policies in rural areas. Yet PF did not serve for any ideological reasons linked with the United States and South Vietnam, out of a sense of nationalism or patriotic zeal fueled by an intense hatred of the VC. They served in their outfits to ensure the well-being of their livelihoods and families.

While Americans in the villages were attempting to overcome cultural and military barriers, the Marines in command of the program dealt with their own obstacles in trying to increase the number of operational

CAPs. In June 1965, when Lt. Gen. Lewis Walt arrived in Da Nang as the commander of the III Marine Amphibious Force (IIIMAF), the designated name for the entire Marine force in Vietnam, he realized that securing and providing for the civilian population should take precedence over large-unit conventional military forays into the unpopulated jungles.[11] Thus, CAPs became an important piece of the Marine Corps' strategy to "win the hearts and minds" of the South Vietnamese from that point forward. However, Walt's approach to the war disagreed with the war of attrition that U.S. Army general William C. Westmoreland implemented as the commander of the U.S. Military Assistance Command, Vietnam (COMUSMACV) from June 1964 to June 1968. As COMUSMACV, Westmoreland enjoyed operational control over all U.S. forces in South Vietnam. Stationed at the MACV headquarters in Saigon, Westmoreland ordered U.S. land-based forces to dispatch large units into the countryside to find and destroy enemy main force units. Many top-ranking Marine officers, such as Lt. Gen. Victor Krulak and Gen. Wallace Greene, accused the army of intentionally undermining the Marine Corps and CAPs. However, solely blaming the army and MACV for the program's failure to blossom at the level the Marines had hoped ignores the overall manpower shortage that afflicted the entire U.S. military in Vietnam. More than the army's war of attrition, the lack of manpower in the IIIMAF area of operations ensured that the program would not flourish at the high level envisioned by the Corps.

In addition to highlighting the differences in strategy between the two American land-based services, this book also details the similarities and differences between CAPs and U.S. Army Special Forces (SF) and mobile advisory teams (MATs). The three had similar characteristics. The American contingents of all lived in or near indigenous populations and trained the local military forces. The Americans in CAPs, SF, and MATs all had cultural adaptations to make during their tours. Yet when one thoroughly dissects each program, various differences also surface. SF and MATs had officers assigned to each unit, and their training was lengthier and more detailed than CAP school. Moreover, the SF had been in operation for nearly a decade before the start of the Vietnam War, whereas the CAP program did not come into existence until August 1965. The emergence of CAPs happened in an ad hoc manner when Marine leaders realized that focusing on enemy rather than civilian bodies would likely not work in Vietnam. MATs emerged in April 1968 as a means to train as many territorial force units as possible before the gradual withdrawal of U.S. forces that commenced in 1969. While MAT soldiers interacted with civilians and local forces, they roamed from one village and territorial force to another, unlike CAP units, which stayed in the same hamlets and worked with their assigned PF platoon for years in some cases.

The specific region of South Vietnam in which IIIMAF operated constitutes one of the primary reasons for the program's failure to grow beyond 114 CAPs. In 1961, under the guidance of the U.S. Military Assistance and Advisory Group (MAAG), the South Vietnamese military divided its country into four corps tactical zones. The RVN appointed general-grade officers in the South Vietnamese army to control these zones. The commander retained operational control of all South Vietnamese military units within his corps tactical zone. Throughout the war, the mainline units of the Marine Corps worked exclusively in the I Corps Tactical Zone, composed of the five northernmost provinces of South Vietnam—Quang Tri, Thua Thien, Quang Nam, Quang Tin, and Quang Ngai (from north to south). All Marine units in I Corps fell under the command of the commanding general of IIIMAF.

The five provinces of I Corps held high strategic value for all belligerents in the war. I Corps extended 225 miles from the demilitarized zone (DMZ) at the 17th parallel to the southern border of Quang Ngai province. The entire corps tactical zone encompassed twenty-eight thousand square miles and contained 2.5 million people, excluding U.S. military personnel. Along the coastal lowlands ran Highway One, the main transportation route in South Vietnam, which stretched from northern I Corps to the Mekong Delta. Eighty percent of I Corps' inhabitants lived among the coastal plains surrounding Highway One, mostly because the region provided fertile ground for rice production, making that area of great importance for the VC foraging for food. Along the western portion of I Corps, bordering Laos, sat Vietnam's Central Highlands, a mountainous and sparsely populated region with dense vegetation and triple canopy jungle. The geography and topography of the Central Highlands proved ideal for North Vietnamese Army (NVA) and VC units seeking to penetrate I Corps via Laos. The Ben Hai River, which stood as the 17th parallel, was the only barrier separating South Vietnam from North Vietnam. Failure to hold I Corps could prove politically and militarily disastrous for the entire U.S. war effort. With all belligerents aware of the strategic importance of northern South Vietnam, I Corps produced some of the bloodiest fighting of the entire war. Examination of the provinces with the highest percentages of total battle deaths during the war shows the five provinces in I Corps ranking among the top ten bloodiest. Considering that South Vietnam had forty-four provinces, this is a staggering figure. The violence in I Corps affected every unit in the field, including CAPs. During the war, 1.5 percent of all the U.S. Marines who ventured to Vietnam served with the program. However, CAPs suffered 3.2 percent of all the U.S. Marine casualties in the war while inflicting 8 percent of total enemy casualties.[12] Despite the lack of manpower and the interservice rivalry, the program actually grew in size every year of the war until the summer of 1969, when it peaked at 114 CAPs. When program manpower climaxed by 1 September 1969, the 1,895 Marines comprised about 2.5 percent of the overall IIIMAF strength in I Corps.[13] While this percentage seems low, one must consider that at its peak, the program had a CAP in about 20 percent of the villages in I Corps.

Every American who served in the Combined Action Program carries with him a unique individual experience. One of the challenges in detailing the Americans' experiences is the heterogeneity of the villages' military and cultural features. The dedication of PF platoons and the generally accepted customs differed from one village to another. One American's experiences in his village differ in many respects from those of someone who served in a different area of operations during the same period. However, one constant for all the Americans in CAP villages remains: each Marine and corpsman had to make the necessary cultural and military adjustments to survive. Many adapted to the environment, while some did not. It is both inconceivable and ignorant to argue that all the thousands of Americans who entered the program departed their villages with a newfound respect for and understanding of the Vietnamese people and their culture. During the war, numerous CAP commanders discarded Marines under their command who had disrespected village customs. Moreover, some CAP veterans today continue to use racist terms to reference the Vietnamese.

One commonality among all the Marines and corpsmen who served in the CAP villages was the unpredictability—often the sheer terror—of their assignments. The Americans were isolated in the rural areas of I

Corps without organic artillery support. With only a squad of U.S. Marines and one corpsman, CAP Americans lived in villages spread over a couple of square miles with civilian populations in the thousands. It was impossible to monitor every person or keep an eye on all corners of the village at all times. Civilians whom the Americans had pinned as friendly often covertly aided the VC. Every day, the small group of Americans knew that at any moment hundreds of enemy soldiers could overwhelm the unit. If a firefight did ensue, the Marines did not know if the dozens of armed PF members would fight with them. At night, some Marines stationed in foxholes and bunkers along the village perimeter never slept out of fear that their PF bunker mate might shoot them in their sleep. Throughout their tours in CAP villages, the Americans operated in an alien cultural environment where communication proved just as worrisome as the military distractions. CAP veterans today are extremely proud of their service, and when they congregate at their annual reunions, one can see the deep emotional attachments they have for each other. Considering what they endured in the war, this comes as no surprise.

Although no one knew it at the time, the Americans in CAP villages were creating a model for future U.S. military counterinsurgency operations. After the terrorist acts on 11 September 2001, the George W. Bush administration wasted no time launching retaliatory strikes against al-Qaeda and Taliban targets in Afghanistan, leading to an ongoing war that by some interpretations has recently surpassed Vietnam as America's longest war. In March 2003, the U.S. military commenced a second war, in Iraq. U.S. combat operations recently ceased there, in 2010. In Iraq, the U.S. Marines implemented CAPs and are currently experimenting with the combined concept in Afghanistan. The U.S. Marine Corps created discussion groups in Quantico, Virginia, bringing together present-day Marines about to embark with a combined action platoon to Iraq and veterans of the Combined Action Program in Vietnam. Only a few CAPs ever existed in Iraq, but they have played a much more prominent role in Afghanistan, where CAP Marines have read accounts of their predecessors in Vietnam in between their everyday duties in Afghan villages.[14]

CHAPTER ONE

The Evolution of
Combined Action Platoons

We had found the key to our main problem—how to fight the war. The struggle was in the rice paddies, in and among the people, not passing through, but living among them, night and day, sharing their victories and defeats, suffering with them if need be, and joining them in steps toward a better life long overdue.

—Lewis Walt

The practice of embedding U.S. Marines among an indigenous population did not originate in Vietnam. The World War I–era Marine Corps first combined the military and political components of a counterinsurgency in Haiti, the Dominican Republic, and Nicaragua. In all three engagements, Marines organized and commanded small units of indigenous military personnel. Alongside the local forces, the Marines conducted patrols against insurgents while providing aid to civilians. The Marines' experiences in these conflicts spawned doctrinal developments in the Marine Corps during the early 1930s that spoke to counterinsurgency and counterguerrilla warfare. However, the later wars against the empire of Japan and the Communists in Korea rendered counterinsurgency doctrine and training obsolete. By the start of the Vietnam War, the Marine Corps had dedicated itself to mirroring the fundamental strategies and tactics of World War II and the Korean War, which had brought the institution military success and the American public's admiration and attention. From the 1930s through the early 1960s, the Marine Corps had developed into America's finest amphibious assault force. In the decades preceding Vietnam, the Marine Corps had paid little attention to the concept of counterinsurgency. Yet in Vietnam, Marine commanders such as Lewis Walt and Victor Krulak quickly realized that the geographic, political, and military landscape of the war was incongruent with the decades of conventional training they had both received and helped to advance. The program emerged spontaneously as part of a larger strategic framework in which Marine commanders sought to gain military and political leverage among the indigenous people in I Corps.

Learning from its experiences in the Caribbean, the Corps formalized a small-wars doctrine in the 1930s.[1] The development of the Fleet Marine Force in 1933, confirming the Marine Corps as an amphibious landing force tied to the U.S. Navy, sparked much controversy within the service over the development of its small-wars doctrine. Many Marines, including the assistant commandant, Lt. Gen. John Russell, envisioned a future Corps prepared for a large conventional confrontation with the world's major powers. Focusing on small wars in peripheral regions would undermine the development and ultimate success of the Fleet Marine Force. Yet the Marine officers advocating the development of small-wars doctrine pushed hard enough to get a manual published in 1935, ultimately titled the *Small Wars Manual* five years later.[2] The 1940 version of the manual defines small wars as "operations undertaken under executive authority, wherein military force is combined with diplomatic pressure in the internal or external affairs of another state whose government is unstable, inadequate, or unsatisfactory for the preservation of life and of such interests as are determined by the foreign policy of our Nation."[3] The manual included detailed descriptions of the military and political components of a small war, much of it based on the Marines' experiences in Latin America. Throughout the late 1930s, Marine schools gradually increased the amount of time allotted for small-war instruction, but lessons never surpassed 10 percent of the overall academic curriculum.

In the two decades following the publication of the *Small Wars Manual,* pacification and counterinsurgency fell by the wayside. In the years preceding World War II, the Marine Corps collaborated with the navy in perfecting the doctrine and practice of the Fleet Marine Force. The Marine Corps hammered its officers with schooling and joint exercises with the navy, learning and practicing the art of launching amphibious assaults to capture forward bases. From 1932 to 1941, Marine officers tackled scenarios in the classroom that tested their ability to devise solutions for assaulting enemy beachheads. "Fleet landing exercises" every winter off the coast of San Diego gave Marines and sailors the opportunity to rehearse the developing Fleet Marine Force doctrine.[4] In World War II, the aggressive, offensive-minded approach to waging a successful war in the Pacific justified the continuance of the Fleet Marine Force concept. The triumphant Marine amphibious assaults against a well-trained Japanese military bolstered the already beaming pride the Corps had in its conventional abilities. The Korean War further boosted the prestige of the Marine Corps. The successful amphibious assault at Inchon in September 1950 and the subsequent annihilation of Chinese forces during the First Marine Division's retreat from the Chosin Reservoir at year's end upheld the Marines' insistence on enhancing the Fleet Marine Force concept. After the Korean War, the Marine Corps continued the development of the Fleet Marine Force, despite President Dwight Eisenhower's "New Look," which emphasized strategic air power and the threat of massive retaliation with nuclear weapons. Facing massive cuts in size and strength in the Eisenhower administration, the Marine Corps attempted to usher in vertical assault to its amphibious doctrine to keep the military branch afloat in the Department of Defense. The advent of helicopters in the Fleet Marine Force of the 1950s also made counterinsurgency seem obsolete. As Marine Corps schooling in the late 1950s opined, "The civil population must be excluded, where possible, from close contact with our forces."[5]

In 1961, President John F. Kennedy began to replace Ike's "New Look" with "Flexible Response," a strategy centered on preparing the U.S. military to fight any type of war, small or large, conventional or unconventional, anywhere in the world. Kennedy's bid to spread the importance of counterinsurgency across all services fell mostly on deaf ears in the Corps. Marine Corps schools implemented counterinsurgency instruction in their curricula, but the Marines learned little about civil affairs, civic action, and population control.[6] In the early 1960s, instruction on counterinsurgency lagged far behind amphibious assault doctrine and division-sized exercises that resembled World War II landings in the Pacific.[7] The commandant of the Corps during the Kennedy administration, Gen. David Shoup, frequently expressed his reluctance to develop counterinsurgency doctrine. Shoup, who became a strong critic of American intervention in Vietnam, deemed counterinsurgency unrealistic and believed, in any case, that it should fall under the auspices of the army, not the amphibious assault-minded Marine Corps.[8] Gen. Wallace Greene, Shoup's successor as commandant, communicated to Congress in 1965 that "the Marine Corps is in the best condition of readiness that I have seen in my thirty-five years of naval service."[9] Greene's analysis of the readiness of the Marine Corps was, in a sense, correct. The Marine Corps was indeed ready to fight a war, but not the type of conflict that erupted in Vietnam. The Marines who landed at Da Nang in March 1965 were products of a Corps that had an ardent affinity for amphibious and vertical assaults, rendering the concept of counterinsurgency insignificant in the minds of some of the highest-ranking and influential Marines. Marine commanders in South Vietnam had to transform a large portion of their conventional force into an unconventional, versatile group that could counter guerrilla activity. Yet when the Marines first arrived for combat purposes in South Vietnam, decades had passed since the Marine Corps had given any serious attention to counterinsurgency.

One of the few exceptions to the Marine Corps' infatuation with conventional war appeared in the 1962 publication of Field Manual (FM)-21, *Operations against Guerrilla Forces.* This field manual acknowledged the successful counterguerrilla campaigns of the U.S. Army in the Philippines at the turn of the century as well as the later British counterinsurgency in Malaya. FM-21 stressed the need in counterguerrilla operations for a temporary patrol base distant from the parent bases. The manual also noted that success in a guerrilla environment depended heavily upon small-unit patrols. Some criteria for a successful counterguerrilla operation in the manual differed from the ultimate mission of CAPs in Vietnam. For example, the manual lays out plans of attack for assaulting guerrilla homes and camps, but CAPs, charged with the primary goal of protecting the villages, did not have as objectives finding and then destroying enemy base camps. Moreover, according to the manual, the best sources of intelligence are maps, recent patrols, photographs, and ground and aerial reconnaissance. Gaining intelligence via interaction with local civilians is conspicuously absent from the intelligence section. Successful counterguerrilla operations, according to the manual, did not entail staying in the villages. Rather, one of the methods for reducing villagers' contact with the enemy was to evacuate or relocate civilians from enemy hotbeds to safer locations, because "areas cleared of civilians provide better areas for tactical operations." The manual does mention the importance of amicable

U.S.-civilian relationships, but it also justifies civilian relocation, arguing that "total or partial evacuation of a given area may be undertaken for the security of the population for imperative military reasons."[10] *Operations against Guerrilla Forces* was a marked deviation from the primary mission of amphibious assaults that so aptly characterized Marine Corps training and doctrine before the Vietnam War. In the same year as the release of FM-21, a U.S. Marine Corps publication found that the service was foremost prepared to commence amphibious assaults.[11] However, storming beachheads in South Vietnam would not earn the respect or win the allegiance of civilians isolated in rural areas.

In Vietnam, the program did not emerge as a direct by-product of Marine Corps doctrine. Instead, the concept of CAPs emerged and progressed in an evolutionary manner, as Marine commanders in 1965 began to realize that the United States could not win the war by relying on its overwhelming firepower and superior technology. When the Marines arrived in Da Nang, they fell under the command of U.S. Army general William Westmoreland, who as commander of MACV held operational control over all American forces in South Vietnam. Just as the Marines in the years leading up to 1965 were obsessed with continuing their amphibious assault doctrine, the army possessed an institutional devotion to conventional war in the form of attrition.[12] Backed by Secretary of Defense Robert McNamara, Westmoreland employed a strategy whereby American forces would use their mobility, firepower, and technology to attain high enemy body counts. With the larger NVA and VC main force units roaming the countryside, Westmoreland saw attrition as the only conceivable choice to win the war. He wanted the Marines to leave the enclaves they had established along the coast of I Corps. Westmoreland and his MACV staff described the enclave strategy as "an inglorious, static use of U.S. forces in overpopulated areas with little chance of direct or immediate impact on the outcome of events."[13] Moreover, Westmoreland feared enclaves would allow the NVA to infiltrate the Central Highlands, rendering the area impenetrable for U.S. forces.

Leading the III Marine Amphibious Force in Vietnam from June 1965 until June 1967, Lt. Gen. Lewis Walt disagreed with Westmoreland's strategy. During his two years as commander of the U.S. Marine force in I Corps, Walt pursued a strategy of protecting the rural population within the enclaves the Marines had established. Walt's inclusion of the Vietnamese people in his strategic blueprint may come as a surprise to some. After all, Walt, who by 1965 had served in the Marine Corps for nearly thirty years, was inextricably connected with a service that took great pride in its amphibious assault capabilities. After briefly serving in the Colorado National Guard, Walt had joined the Marine Corps in 1936 as a second lieutenant. By the fall of 1942, after his service with the First Marine Raider Battalion on the Solomon Islands and the Fifth Marines on Guadalcanal, Walt had been promoted to major. He went on to lead Marines at Cape Gloucester and Peleliu, ultimately returning to the United States in November 1944, having earned the Silver Star and two Navy Crosses for his participation in the Pacific. Walt commanded the Fifth Marine Regiment in the Korean War and later accepted positions in Quantico, Virginia, as an educator of Marine officers. Just before landing the job as IIIMAF commander, Walt had directed the Marine Corps Landing Force Development Center.

While Walt's accomplishments in the 1940s and 1950s certainly bolstered his stature as a U.S. Marine officer, they do not shed light on why he sought to protect the rural population of I Corps and in turn become one of the chief architects of the Combined Action Program. Indeed, counterguerrilla and counterinsurgency operations seldom occurred in the Pacific theater of World War II or in the Korean War. More than any other era in his illustrious career, Walt's pre–World War II service provided the future IIIMAF commander with a general knowledge of "small wars" and the importance of aiding the civilian population. In 1936, as a newly commissioned junior officer, Walt attended the Basic School in Philadelphia. In the 1920s and 1930s, the school conducted classes on the "small-war" experiences of the Marine Corps in Haiti, the Dominican Republic, and Nicaragua. Capt. Lewis "Chesty" Puller was the primary instructor of Walt's group of junior officers.[14] Puller used the *Small Wars Manual* as a general text, supplementing it with his personal wartime experiences in Haiti and Nicaragua. His recollections taught the junior officers the values of living off the land and engaging with the indigenous population in a productive way. According to one of Puller's biographers, Walt "absorbed every word" that "Chesty" said.[15]

When Walt took control of IIIMAF in the spring of 1965, he admitted that the Marines had many strategic challenges ahead of them in I Corps. Walt faced the grim reality that the war in Vietnam would not feature many of the amphibious assaults that had symbolized the Marine Corps and its doctrine in the previous decades. Gradually, as the Marines opened three separate enclaves in I Corps during 1965, spreading from Thua Thien province to Quang Ngai province, Walt recognized, "It was a new kind of war we were in, where concern for the people was as essential to the battle as guns or ammunition, where restraint was as necessary as food or water."[16] From the spring of 1965 until their departure six years later, the Marines created their own way of fighting the war in Vietnam. Although adhering in part to MACV's attrition-based strategy, the Marines exuded an unmatched dedication to securing and controlling the rural population.

Engagement in civic action was vital for creating amicable relations between the Americans and the Vietnamese. Early civic action in I Corps usually consisted of infrequent medical aid from corpsmen along with the distribution of C rations and soap to civilians. Throughout South Vietnam, government agencies and nongovernment organizations (NGOs) that provided supplies for civic action worked independently of each other. The result was disorganization and a time-consuming bureaucratic maze through which supplies were channeled before reaching their destination. Frequently, agencies failed to communicate with each other, resulting in lost supplies. Poor coordination also resulted in inequities: an abundance of supplies might land in one rural area while other villages received none.

In 1965, Walt amalgamated the multiple civic action agencies into the Joint Coordinating Council (JCC), a group of military and civilian representatives who collectively managed and organized the pacification effort in I Corps. The JCC determined requirements for agencies aiding Vietnamese civilians in I Corps and recommended specific procedures to make the multiagency process more efficient. Originally consisting of seven members, including representatives from IIIMAF, the MACV, the U.S. Operations Mission to Vietnam (USOM), and the Joint U.S. Public Affairs Office (JUSPAO), the JCC formed committees in special areas of interest for pacification and civic action such as public health, education, logistics, and finance. The JCC remained mindful of Saigon's instructions for the overall pacification effort in South Vietnam, consistently molding its overall mission according to the RVN's plans for rural development.

Also beginning in 1965, Walt oversaw various Marine programs to help pacify villages within the enclaves. Before the Marines' arrival in

I Corps in 1965, the VC had frequented villages to enforce rice taxes during harvest season. According to Walt, the rice demands from the VC varied from 25 to 90 percent of a village's total harvest.[17] To keep the rice in the hands of the villagers, IIIMAF launched "Golden Fleece" operations, in which Marines would cordon off hamlets to keep the food out of VC hands. Meanwhile, the Marines transported rice stocks to secure locations outside the villages until the end of harvest season. U.S. helicopters airlifted the villagers to the rice stocks and returned them the same day with the needed allotment of food. With two rice harvests per year, the Marines could plan when to begin and end Golden Fleece operations.

In 1966, the Marines began "County Fair" operations throughout their enclaves. A company of Marines provided security around hamlets while Army of the Republic of Vietnam (ARVN) units and South Vietnamese police set up an assembly area inside the perimeter where they conducted a census, issued identification cards, questioned villagers about recent VC activity, and determined the whereabouts of missing civilians. During the daylong operation, Marines and ARVN soldiers offered medical treatment to villagers, while those waiting their turn watched propaganda films and listened to popular Vietnamese songs blaring from speakers. Perhaps the most gratifying component of the County Fair operation for the villagers was the massive amount of food and cold drinks handed out.[18] Although South Vietnamese territorial forces replaced the Marines after a County Fair operation, these soldiers were poorly trained, leading to undermined security in the absence of a continued American military presence. By 1967, with IIIMAF resources stretched thin from the relentless NVA attacks in northern I Corps, the number of personnel dedicated to County Fairs began to steadily decline in favor of line units in the field. Although the program was short lived, County Fairs revealed to IIIMAF that for pacification and civic action to thrive, a permanent and effective military presence in the villages would be imperative in reducing the civilians' contact with the VC. The Fleet Marine Force, Pacific (FMFPAC) commander, Lt. Gen. Victor Krulak, who was in charge of training and supplying Marines in Southeast Asia, believed the Marines provided the solution for solidifying village security. By the spring of 1966, the FMFPAC commander had all but given up on the South Vietnamese village militias' ability to defend themselves and the rural infrastructures. "Obviously, they cannot do these things themselves," Krulak wrote. "They need our help."[19]

The earliest instance of combining Marines with PF occurred in May 1965 in the village of Le My, northwest of Da Nang. After several Marine patrols had dodged enemy sniper fire near Le My, Lt. Col. David Clement, the commander of the Second Battalion, Third Marine Regiment, ordered his infantry to clear the village of the apparent VC presence. That same month, the Marines succeeded in securing control of Le My, upon which they began patrolling the area over the next few days. Ultimately, the district chief agreed to dispatch a PF platoon to relieve the Marines at Le My. Yet the inexperience of the PF forced the Marines' return to the village to train the South Vietnamese force in ambush tactics. The Marines were also needed to render medical aid to civilians. VC harassment in and near the village decreased over time, but it was evident that the village would not remain secure without the Marines' continued presence. Walt knew that in order to ultimately defeat the VC, the Marines also had to demonstrate sympathy toward and understanding of the civilians. To achieve this, Walt presumed, "required a gentler touch."[20]

In August 1965, on the heels of the Le My experiment, Marines from the Third Battalion, Fourth Marine Regiment erected the first CAP near Phu Bai, just south of the Vietnamese imperial capital of Hue. The battalion had landed in April 1965 to protect the Phu Bai airstrip in the center of the enclave. The VC had been operating freely along the fringes of the Phu Bai enclave, which encompassed only two square miles. The Third Battalion commander sought to extend the enclave to include three nearby villages that were within enemy mortar range of the airfield. In June, the Marines succeeded in this endeavor, allowing the battalion to move into the villages that had caused concern. Marine rifle platoons secured the perimeter of the villages, while civil affairs teams distributed medical, food, and sanitary supplies to the villagers. At first, mutual mistrust and misunderstanding kept both the Americans and the South Vietnamese cautious in their approaches to each other. The first major stepping-stone in constructing a more comfortable social environment came over the next week when the battalion commander met with each village's political council. The Vietnamese constituents agreed to accept weekly American medical services in each village, which the civilians gradually began to embrace.[21] As the medical aid gained popularity, Marine "grunts" began to coordinate combined patrols with PF soldiers around the villages. Yet the obvious inexperience of the PF worried the Marines, who often patrolled without the local military force.

Intelligence from the villagers about VC activity was critical for the units to succeed, but the lackluster security offered by the PF in the Marines' absence made villagers reluctant to divulge information. With the village ineffectively protected, the VC had easier access to civilians. The villagers knew that communicating with the Americans increased their risk of violent reprisal from the nearby VC. Capt. John Mullen, who served with the Third Battalion, Fourth Marine Regiment at the time, described the Marines' dilemma: "Although we had gone about our civil affairs program according to the 'book,' we had neglected the very important factor of security for the population. Thus, our effort so far had gained absolutely nothing."[22] When the Marines and village councils convened for a second time, the South Vietnamese inquired about the possibility of having the Marines stay in the villages to bolster security. Acceding to the villagers' request, the battalion commander agreed to assign Marines to the villages on a permanent basis. The Americans would be responsible for security, civic action, establishing a civil-military relationship, and developing an intelligence network. The plan centered on attaching a small unit of Marines to each village's PF platoon to coordinate security efforts in and around the villages. The battalion commander sent only a squad of Marines to each village because IIIMAF possessed neither the manpower nor the resources to justify moving larger units away from the major urban areas, the political and military centers of the enclaves. In late July 1965, the battalion commander obtained approval from his superiors to combine the Marines and PF together to collectively perform military security and civic action. On 1 August 1965, the first CAP was born.

First Lt. Paul Ek, attached to the Third Battalion, Fourth Marines as an interpreter, became the commander of the first CAP, then called a Joint Action Company. Ek and Mullen selected the Marines who would spearhead the new combined units, planned to operate in four villages with fourteen thousand people. In early August, the Marines began conducting daylight patrols near the villages, and each American contingent acquainted itself with its assigned PF unit.

In September, the Joint Action squads gradually increased time spent in the villages, rising to three to four nights per week. Mullen replaced

Ek as commander in late September, a month that also saw the Marines' initial contact with the VC and the first CAP Marine killed in action (KIA). By November 1965, 75 percent of all combat operations in the combined units stemmed from tips offered by the villagers.[23] As the year ended, many of the Marines in combined units had voluntarily extended their tours for six months rather than return to Okinawa with their parent battalions.[24]

In October, IIIMAF replaced the term joint action company with combined action company. The name change made sense, since a joint military effort consists of troops of multiple military branches from the same nation working together, whereas a combined unit is comprised of a mix of soldiers from different countries. By October, six combined action companies were in operation under the command of their parent mainline Marine battalions. Each parent battalion then supplied one corpsman to each combined company to offer convenient medical attention to the Marines, PF, and civilians.

In November 1965, Walt asked the commanding general of the Third Marine Division to assess how his subordinate units were using the PF in their areas of operation. The Third Marine Division's headquarters informed Walt of its desire to create a closer military relationship with the PF. Walt wanted to expand the number of combined units, but he first needed approval from the South Vietnamese commanding general of I Corps, Maj. Gen. Nguyen Chanh Thi. After the PF from the newly created CAPs in Da Nang had shown progress, Thi authorized the expansion of the program to all Marine enclaves in I Corps. By December 1966, the program had grown to fifty-eight CAPs, a marked increase from the six CAPs in existence just one year before.

The decision to place a CAP in a village depended on several factors. The IIIMAF commander and his South Vietnamese counterpart in I Corps had to come to a mutual agreement before activating a new CAP. Then, at least part of the target village had to show some semblance of RVN control, and the location had to be accessible to roads for supply purposes and within range of the nearest air and artillery units. Moreover, the combined U.S.-RVN command structure in I Corps had to deem the PF of a village sufficiently inadequate to justify displacing Marines from mainline units for CAP duty. Sending Marines for months and possibly years to train a PF platoon with a history of effectiveness and efficiency did not make sense. After the district chief of the proposed CAP village approved, the first Marines and corpsmen arrived in their new area of operations.

The year 1967 proved to be a pivotal turning point for the program. Until July, regular U.S. line units held operational control over CAPs within their respective tactical areas of responsibility. The local U.S. infantry battalion was also responsible for logistics, fire support, and reaction forces for the CAPs. In the program's first two years of existence, CAPs often performed military functions instead of their given tasks at the request of the parent unit outside the village. For example, the Fifth and Seventh Marine regiments used their CAPs on three-day sweeping operations, leaving the villages unprotected.[25] In June, Gen. Robert Cushman replaced Walt as IIIMAF commander. By the time of Cushman's assignment, the program had an official table of organization and equipment (TOE), giving IIIMAF headquarters direct operational control of the CAPs. This was a watershed moment for the program because it solidified the new standard operating procedure for CAPs, removing the line units from the program's chain of command. The program no longer had to worry about U.S. military units in the field neglecting the pacification and counterinsurgency duties of the CAPs under their command. Cushman placed his deputy IIIMAF commander, Maj. Gen. Herman Nickerson, in charge of the program. Nickerson, who had previously commanded the First Marine Division in Vietnam, then delegated most of the authority to the new director of the program, Lt. Col. William Corson. Before his new assignment, Corson had commanded a Marine tank battalion that successfully pacified the village of Phong Bac in Quang Nam province.[26] As the new director in 1967, Corson quickly worked to formulate a standard operating procedure for the program, outlining its exact missions, goals, and chain of command. Corson also created a school in Da Nang to teach Marines small-unit tactics and the Vietnamese language and culture and formulated the criteria program applicants had to meet before joining a CAP. To bring more organization and structure to the program, Corson initiated combined action groups (CAGs) at the province level. With the exception of the First CAG, which was responsible for the two provinces of Quang Ngai and Quang Tin, each province in I Corps ultimately housed one group compound. With four CAGs in operation by July 1968, each was responsible for several combined action companies (CACOs) at the district level, which in turn monitored the numerous CAPs in the villages.[27] By December 1967, when the number of CAPs had not grown at the rate Corson had hoped, he left the program director position and returned to the United States, where he published a scathing critique of military strategy in the war.[28] Corson had already made his mark on the program, but the exact reasons for his departure, besides the frustrations over MACV strategy, remain unclear.

Although the program experienced growth in numbers and in administrative organization by the summer of 1967, the continuing menace of NVA units near the DMZ halted the rapid surge of CAPs. Beginning in early 1966, the NVA had begun to harass U.S. and ARVN units in the two northernmost provinces of I Corps. Until that year, Walt had widely dispersed Marine units across I Corps. As the NVA increased its attacks near the DMZ in 1967, Westmoreland dispatched U.S. Army units to reinforce southern I Corps, and Walt moved Marine mainline units from southern I Corps to the northern provinces to assist in repelling the Communist assaults.

At the beginning of 1967, the commander in chief, Pacific (CINCPAC), Adm. Ulysses S. Grant Sharp, and MACV headquarters articulated goals for IIIMAF. With hopes of achieving high enemy body counts and neutralizing enemy base areas and logistical routes, Sharp and Westmoreland wanted IIIMAF to increase the size of the program to 114 CAPs by the end of the year. Yet as the NVA increased its attacks near the DMZ, the idea of taking personnel away from the line units and placing them in CAPs could prove disastrous for IIIMAF forces already stretched thin across I Corps. The number of CAPs remained at 75 in May, June, and July 1967. However, every other month saw an increase. While the number of CAPs in 1967 jumped from 57 in January to 80 in November, the total number of Marines in I Corps fell by more than 2,000 during that same period, mostly due to the increasing army presence. By December 1967, 1,328 of the 71,242 Marines in I Corps worked in the program at the platoon, company, and group levels.[29] Although the program did not achieve its projected goal of 114 CAPs by the end of 1967, its growth, considering the volatile military situation in I Corps that year, is noteworthy. The increasing number of CAPs in 1967 stemmed from the growing number of army soldiers in I Corps but also was a testament to the high regard that IIIMAF had for the program.

The year 1967 also saw the creation of Civil Operations, Revolution-ary Development Support (CORDS), which placed all pacification activi-ties under MACV auspices, with a deputy civilian given equal command and control. As the inaugural civilian leader of CORDS, Robert "Blow-torch" Komer carried the status of ambassador and had U.S. military supplies and logistical resources at his disposal to help complete the paci-fication goals that he set. Every province and district of South Vietnam ultimately possessed CORDS representatives who implemented Komer's policies. The primary concern for CORDS throughout the rest of its ten-ure rested with improving the territorial forces, a prospect that alarmed IIIMAF headquarters with the possibility of losing CAP PF.

The Marines were concerned that CORDS would attempt to dimin-ish or wrangle complete control of their pacification efforts in I Corps. When CORDS identified the specific provinces it believed should receive priority over others, only one province (Quang Ngai) in I Corps made the list. CORDS placed a strong emphasis on III Corps and IV Corps to the south. The priority provinces constituted areas that had shown progress in pacification and a capacity to achieve greater results in the future. During the fall of 1967, Lt. Gen. Wallace Greene, then the commandant of the Marine Corps, wondered why the IIIMAF area of operations was largely overlooked by CORDS when "only in the IIIMAF area of responsibility has the target of pacification, civic action, and revolutionary development been accorded primary emphasis from the outset of U.S. major involve-ment in Vietnam."[30]

CORDS did not have operational control over the program, but it did regulate the territorial forces and the IIIMAF civic action program. CORDS did, however, attempt to control CAPs and use them in ways that undermined the standard operating procedure of the program. For example, shortly after the creation of CORDS, the senior province advi-sor of Quang Tri tried to extract a CAP from its village to assist a Re-gional Forces company in the CAPs' district, which would have taken the Marines away from their confined area of operations.[31]

Although Komer deemed the program a success in early 1968, the CORDS director suggested that CAPs change their tactics to mirror U.S. Army mobile advisory teams (MATs). Komer wanted CAPs to split into smaller, eight-man teams to increase the number of hamlets occupied by Marines. He also proposed moving CAPs out of their villages at a much more rapid pace.[32] Both of Komer's proposals would have completely changed the standard operating procedure of CAPs, distancing the Marines from the civilians who provided crucial intelligence. IIIMAF and the program were not willing to accept this change.

In February 1966, the JCC created its own system of gauging how IIIMAF would measure the success of Saigon's pacification plans, which differed from the hamlet evaluation system (HES) created by CORDS. The Marine system for measuring pacification success depended on five variables: destruction of enemy military units, destruction of enemy infrastructure, establishment of security by local South Vietnamese, establishment of a local government by the South Vietnamese people, and the status of new life development programs. Each of the five categories was given a score from 1 to 20. Thus, the maximum score (and a perfect one) was 100 points. A score of 60 for a village meant there was "significant RVN/U.S. influence." CORDS monitored the progress of its pacification efforts by assigning to each hamlet a letter grade ranging from "A" (under firm RVN control) to "E" (contested hamlet). A "V" signified a hamlet under VC control.

In addition to the differences between the numerical and alphabetical means, the IIIMAF system gauged village security, whereas the HES evaluated pacification progress at the hamlet level. The HES entrusted district advisors with visiting hamlets, but the CORDS personnel administering the reports did not have time to gauge progress in all of the villages in their assigned area. The Marines used their own military structure to obtain information, and CAP commanders in the villages usually completed the reports. The overriding difference between the two systems was that the HES looked for improvements in administrative, political, and economic factors in the hamlets more than military conditions. For the Marines, it was the complete opposite.

In 1969, IIIMAF adopted HES as the standard rating system for measuring pacification progress in I Corps. This was due in part to the RVN's 1969 pacification and development plan, based solely on HES ratings. Although I Corps' ratings decreased under the new HES system, the reports showed that in early 1970 more than 85 percent of I Corps' 2.5 million residents lived in villages considered under U.S./RVN control.[33]

During the Tet Offensive of 1968, the VC (and to a lesser extent the NVA) targeted all major urban areas in South Vietnam. With the larger urban areas the target, CAPs played a peripheral role in the rural villages relative to the sustained fighting in the city streets of South Vietnam. Yet many CAPs did experience an increase in contact with their enemy. From the two days prior to the offensive (29–30 January) until the end of the first week of February, the number of CAP contacts with the VC tripled. With the increased activity came more casualties for CAP units. December 1967 saw Marines and PF in CAPs suffer 8 killed and 25 wounded. During January 1968, those numbers jumped to 65 killed and 125 wounded.[34] Cushman, IIIMAF commander during the Tet Offensive, praised CAPs for delaying the enemy's attack on the cities of I Corps, thus leaving the cities' defenses better prepared.

During the attacks, sixty-nine of the eighty-nine CAPs remained in their villages, while the others received orders to relocate to provide security at critical transportation junctions such as Highway One between Da Nang and Phu Bai. Before the offensive, villagers from one CAP had begun abandoning their homes, which signified to the Marines that something was about to happen. Another visible hint came from the villagers in Igor Bobrowsky's CAP, who began constructing an unusually large number of coffins. Bobrowsky recalls thinking, "What the hell are these people making coffins for? Who died? Was there, you know, a plague?"[35] Some CAP Marines sensed an abnormally large amount of enemy movements near their areas of operation. One reported to his superiors the movement of VC heavy weapons squads near his CAP village. Yet, according to the CAP Marine, "Nobody would bother to check it out."[36] As far as the big picture went, CAPs did not know that an onslaught as large as the Tet Offensive was looming.

The massive VC movements across I Corps alerted the program to the vulnerability of isolated, stationary CAP units trying to fend off overwhelming numbers of enemy forces. Before the Tet Offensive, the VC had begun to memorize the predictable routines and repetitive patrol routes of the "compound CAPs"—those that had fixed compounds, or forts, at a central location in the villages. Compound CAP units tended to congregate and plan patrols near the fortified structures. During the first three years of the program, all the combined units were compound CAPs. In the wake of the Tet Offensive, all existing and newly created CAPs gradually adopted the mobile concept, in which the Marines and PF became guerrillas within their own area of operations. "Mobile CAPs" featured unpredictable patrols that traveled from one location to another outside the village perimeters, never remaining in the same place for more than a few hours, in hopes of avoiding routines easily detected by the VC. The mobile concept inherently sacrificed interaction with the villagers for more efficient and effective military operations. With the Marines constantly on the move, they had less time to communicate with civilians and assist in large civic action construction projects.

Statistics show that mobile CAPs suffered fewer casualties and killed more of the enemy than the compound units.[37] However, figures alone quite naturally cannot assess wholly the changing dynamics of the war after the Tet Offensive. The VC bore the brunt of the fighting against American and South Vietnamese forces. Out of an attacking force totaling approximately eighty thousand soldiers, the NVA and VC sustained somewhere between thirty thousand and sixty thousand casualties, the bulk of which were among the ranks of the Communist-aided insurgents in the south—the primary enemy of CAPs.[38] The VC never regained the effectiveness or the size in manpower that it possessed before the offensive. The official history of the People's Army of Vietnam describes the latter half of 1968 as a time in which the "political and military struggle in the rural areas declined and our liberated areas shrank."[39] The implementation of the mobile concept may have increased the effectiveness of CAPs statistically, but the military situation in I Corps had changed rather dramatically regarding the strength of the VC.

In the midst of the VC's decline in numbers, the program increased in size, at least for a time. Beginning in 1969 and into 1970, the program peaked at 114 CAPs, but the apex occurred precisely when the United States began withdrawing troops from South Vietnam under Richard Nixon's Vietnamization policy. CAPs dissolved at a pace equaling that of the overall departure of all Marines in IIIMAF. After all, CAPs relied on conventional Marine units for artillery and air support as well as for infantry reinforcements. In 1969, IIIMAF commander Nickerson ensured close supervision of the deactivation of CAPs. Nickerson feared that a rapid decrease in CAPs could result in disaster for the villages.[40] Beginning in 1970, the 114 CAPs in I Corps slowly began to deactivate. The program then placed its focus on CAPs in Quang Nam province. By 1 September, all CAPs outside Quang Nam had been deactivated.

By April 1970, a "relative calm" in I Corps allowed the CAPs to focus more on training the PF in preparation for the Marines' inevitable departure over the coming months. In May, the program began to deactivate CAPs, and by September, only thirty-four remained in Quang Nam province. The entire program ended in May 1971.[41] Gen. Ngo Quang Truong, the commander of the First ARVN Division, pleaded with the Marines not to remove the CAPs. "I don't care what else you do," Truong exclaimed, "but please don't take the CAPs."[42]

In March 1970, the U.S. Army's XXIV Corps, as the senior U.S. command in I Corps, assumed control of the program. The XXIV commander, Lt. Gen. Melvin Zais, assured the Marines that he would continue to manage the program as IIIMAF had done for the previous five years. By the time XXIV Corps took control of I Corps, Vietnamization had taken hold and the fight from IIIMAF headquarters to retain command of CAPs had subsided significantly.

With the eventual reduction of CAPs under Vietnamization, IIIMAF devised new ways to keep an American military presence in I Corps villages. In late 1969, the Infantry Company Intensified Pacification Program (ICIPP) emerged as an outgrowth of the Combined Action Program. Run by the First Marine Division and the U.S. Army's American Division, the ICIPP combined infantry rifle companies and corpsmen with PF. CAPs still existed, but the IIIMAF Marines and army personnel assigned to the ICIPP came directly from infantry units. These soldiers did not receive the training the CAP Marines had, nor did they need to meet particular criteria. In 1970, the ICIPP became known as the Combined Unit Pacification Program (CUPP), and at the beginning of the year it operated in twenty-six hamlets. CUPPs existed until April 1971. Peaking at only twenty-four, some of which monitored multiple hamlets, CUPPs never came close to attaining the number the CAPs reached.

Under CUPP, territorial forces combined with American infantry units at the company, platoon, and squad levels. As in CAPs, neither the South Vietnamese nor the Americans had operational control over the other, but they did cooperate in formulating patrol and ambush plans. Each American infantry unit had one territorial forces unit to train and a specific area of operations that equaled the size of the CAPs' areas of responsibility. Although the tenure of the CUPP organization was relatively brief compared to that of the CAPs, time spent in CUPP villages equally changed the attitudes of many of Marines toward the Vietnamese.

The doctrinal developments and training regimen of the Marine Corps in the decades preceding the Vietnam War had built upon the amphibious assault tradition that had brought the Corps campaign victories and highly desired publicity. The Marine Corps did not score victories at Iwo Jima and Inchon because the Marines involved actively engaged civilian populations. Indeed, preparing Marines to live alongside foreign civilian populations in their homes fell far outside the institution's amphibious assault traditions. When Lew Walt arrived in Vietnam as the IIIMAF commander, he had not designed any kind of counterinsurgency blueprint for the war. Nor did he have a field manual that provided guidelines for how to send Marines into villages to live with civilians. The creation of the program was spontaneous. Walt had quickly identified a lapse in the U.S. military's strategic approach in Vietnam, and through the Combined Action Program he endeavored to correct that lapse.

As the Combined Action Program gained popularity and strength in I Corps, various colonels and generals in the U.S. Army disliked the Marines' way of war in Vietnam. However, the army did not completely ignore pacification and counterinsurgency. In addition to participating in the CUPP program alongside the Marines in I Corps, the army had its own unconventional units sprinkled across South Vietnam. Although similar to CAPs in that they lived near and interacted with civilian populations, the army's counterinsurgency and pacification units had numerous technical differences when compared with the program.

CHAPTER TWO

Combined Action Platoons, Green Berets, and Mobile Advisory Teams

Our men have learned to live with them, learn their language and patois
and customs and taboos and to win their confidence and respect.
—Brig. Gen. William P. Yarborough, U.S. Army

On the surface, U.S. Army Special Forces A-teams and mobile advisory
teams shared characteristics with CAPs. SF A-teams and MATs lived near
the civilian population, interacted with villagers, instituted civic action,
and trained the local indigenous military forces. Living in or near South
Vietnamese villages guarded by inexperienced local forces, American soldiers in SF and MATs had to overcome the same general military and
cultural barriers as CAP Marines. Yet digging below the surface of SF,
MATs, and CAPs reveals numerous differences between the three, most
notably in the arenas of length of tenure in Vietnam, the overall purpose
of the units, their training, the military composition and ranks of the soldiers versus the Marines, and the cultural and ethnic backgrounds of the
villagers whom the Americans worked alongside. SF, MATs, and CAPs
each had their own share of success in training the indigenous forces and
creating a rapport with the civilians, but a detailed comparison of the effectiveness of the three is beyond the scope of this project.

The beginnings of the three programs in Vietnam occurred at different transitional stages of American involvement in Southeast Asia. SF first
arrived in South Vietnam in the early 1960s during the U.S. advisory period, remaining in place during the subsequent buildup and withdrawal of
American combat forces. SF A-teams, composed of twelve soldiers, lived

with and trained montagnards, a collection of ethnically diverse tribes in the Central Highlands.[1] A-teams stayed in South Vietnam during the course of U.S. military involvement in the war, but MACV changed the nature of SF's operations from counterinsurgency to border surveillance in the Central Highlands. As outlined in the previous chapter, the emergence of CAPs in 1965 coincided with the arrival of America's first combat troops to South Vietnam. Although the MACV commander held operational control over the Marines, Westmoreland and his successor, Gen. Creighton Abrams, gave the Marines leeway in their handling of the program. IIIMAF never changed the general structure and purpose of CAPs. The switch from compound to mobile units changed the tactics used in the villages, but the counterinsurgency template for CAPs held true throughout the war. After the Tet Offensive, which heightened political upheaval in the United States and forced the U.S. military to reexamine its strategy, CORDS began focusing on training as many territorial force units as possible to prepare them for war without American assistance. To achieve that goal, in April 1968 CORDS and MACV created MATs, five-man teams of roaming U.S. Army advisors that jumped from one territorial force unit to another within a district, training the local forces and providing security for the nearby civilians. By the fall of 1969, when U.S. forces had begun to withdraw from South Vietnam, MAT advisors became the American military's primary instruments for training territorial force units across all four corps tactical zones.

Army SF, conceived in the early 1950s, garnered more attention from the Department of Defense when President John F. Kennedy ushered in his "Flexible Response" strategy, which urged the U.S. military to strengthen its unconventional abilities during the cold war.[2] Green Berets, the name given to SF soldiers during the Kennedy presidency in a reference to their military headdress, were some of the most elusive and well-trained warriors in the entire U.S. military. Army SF in the early 1960s were prepared to fight unconventional wars anywhere in the world. By the time the U.S. Marines arrived in Vietnam in 1965, the SF had already occupied territory in the most remote areas of the Central Highlands for several years. In fact, an SF officer earned the first Medal of Honor in the Vietnam War for his heroics eight months before the Marines landed in Da Nang.[3]

The standard SF unit that operated in South Vietnamese villages was the A-team. Each A-team had the same number of troops as a standard

CAP; however, it had two officers, whereas the program dispatched enlisted Marines and NCOs in the villages. First lieutenants filled the two officer ranks of an A-team, one as the executive officer and the other as the psychological warfare operations officer. Rounding out the remaining ten spots of an A-team were enlisted specialists, each trained in intelligence, weapons, medicine, demolitions, or communications. Every member of an SF A-team had cross-trained in a secondary field; each could perform secondary specialties in addition to his primary function. With two officers and ten enlisted soldiers, the cross-training increased the versatility of the SF units, allowing A-teams to split into two separate six-man groups if needed. Although this meant that the medical specialist tended to only one-half of the A-team, the other six-man contingent possessed a Green Beret who had cross-trained in the medical field.

A-teams in Vietnam had additional SF soldiers providing reconnaissance and serving as a quick reaction force for the isolated Green Beret compounds. Mobile strike forces, known as "Mike" teams, further augmented the A-teams by serving as their forward observers. Mike forces consisted of SF soldiers who worked with a plethora of ethnic groups, including montagnards, Chinese Nungs, Vietnamese, and Cambodians. In addition to Mike teams, A-teams received supplemental support from SF mobile guerrilla teams, which spent up to sixty days in the jungles of South Vietnam, operating independently of any other American or South Vietnamese unit. Their missions primarily centered on disrupting VC strongholds by creating forward ambush sites and calling in air strikes.[4] The use of Mike teams and mobile guerrilla units highlights another difference between the SF and CAPs. The only quick reaction force for CAP Marines came from either nearby territorial force units, usually the closest Regional Force (RF) contingent, or the nearest American mainline unit. In the same vein, the source of intelligence for CAPs came from daily and nightly patrols or from the villagers. The army strike forces aspired to extend the perimeter of one secure village to include others, much like the Marine enclave strategy, but CAP patrols rarely covered ground outside their given area of operations around the village. Some I Corps districts did have "double CAPs" that featured at least two combined units working together, but these were exceptions for the program.

SF base camps proved more extensive than those in CAPs. Because of their isolation in the Central Highlands away from major ground transportation routes and out of range of artillery support, A-team base camps often featured a landing pad for helicopters or an airstrip capable of handling C-123 cargo planes that offered daily supply runs for the soldiers. Roads connecting SF camps to larger towns were virtually nonexistent. In contrast, most CAPs rested on the outskirts of Marine enclaves, relatively close to major roadways. The villagers and the Marines had access to regional markets in the lowlands surrounding Highway One. In the more densely populated lowland regions of I Corps, CAP villages interacted with outside visitors looking to trade or sell products. The most frequent visitors to SF camps were either South Vietnamese or American political and military officials—or the enemy.

The SF camps had two main buildings, one quarters for the Americans, Vietnamese SF, and interpreters, the other quarters for the montagnard soldiers. Some SF teams divided the buildings along nationality lines, with the Americans in one building and the South Vietnamese in the other. Smaller buildings were scattered around the main base camp, including a mess hall, a communications bunker, a dispensary, and living quarters for nurses. Amid the cluster of buildings within the barbed wire sat montagnard civilian houses. SF frequently removed montagnards from their villages to makeshift homes near the main base camps. The compounds constructed by Marines in CAP villages were heavily fortified and contained bunks for sleeping, but they did not feature a vast array of smaller military buildings surrounding them. The only structures near CAP compounds were villagers' homes.

An analysis of the history of the strategic employment of SF in the Vietnam War helps one understand how the colonels and generals in the army differed from their Marine counterparts in implementing counterinsurgency. In 1961 the U.S. Central Intelligence Agency (CIA), which then controlled the SF, created the Civilian Irregular Defense Group (CIDG) program, arming montagnards in the Central Highlands to secure their villages under the supervision of Green Berets. The SF soldiers assisted the montagnards in fortifying villages and military buildings within the barbed-wire confines. By the early 1960s, the VC had begun recruiting montagnards, attempting to take advantage of the historical rift between the Central Highland tribes and the South Vietnamese government. Recognizing the strategic importance of the Central Highlands, which provided logistical routes for NVA and VC forces operating in or between

Laos, Cambodia, and South Vietnam, the U.S. Mission in Saigon sought to implant A-teams into montagnard villages. The Central Highlands provided an excellent buffer zone for the military force that could gain control of the mountainous region and its population. Gen. Vo Nguyen Giap, the chief architect of North Vietnam's military strategy, once argued, "To seize and control the highlands is to solve the problem of South Vietnam."[5] By the spring of 1962, CIDGs had succeeded in pacifying forty villages in the program's initial target province of Darlac. By July, the CIA had requested more SF soldiers to spread the pacified "inkblot" in the Central Highlands.

Despite the initial gains, MAAG and ultimately MACV disliked the unconventional nature of the CIA-controlled CIDG program. U.S. Army generals William Rosson and Maxwell Taylor believed the CIA had improperly used the CIDGs and SF units. Army Chief of Staff Gen. Harold Johnson criticized the SF for building fortifications out of the "Middle Ages" to "bury themselves . . . with concrete."[6] In other words, the army wanted the SF to partake in offensive operations. In 1963, Operation Switchback transferred control of the CIDG program from the CIA to the Department of Defense and ultimately MACV. With full operational control, MACV transformed the CIDG Program into an offensive weapon for the war of attrition, using the indigenous units as mobile strike forces. MACV rapidly increased the size of the CIDG Program, but a shortage of Green Berets forced U.S. Army commanders to rely more on the Vietnamese Special Forces, which had exuded poor leadership and racial hostility toward the montagnards.[7] By 1965, MACV had changed the primary mission for SF from counterinsurgency to surveillance along the Laos and Cambodia borders. Westmoreland deemed the tactical concepts of SF inadequate because of their emphasis on defense rather than on aggressively pursuing guerrillas outside their bases.[8] As Andrew Krepinevich argues, the MACV takeover of SF operations solidified them as an unconventional force, rather than one centered on counterinsurgency. Unconventional operations, the organizing of partisan and guerrilla forces to harass the enemy, stood in stark contrast to counterinsurgency, with its primary aim of defeating the insurgent infrastructure. According to Krepinevich, the unconventional nature of the SF played into the hands of MACV's desire to use them and the entire U.S. military in traditional, conventional operations.[9]

The military takeover of the CIDG Program transported indigenous soldiers and their families to separate camps, as the combined units searched for larger VC forces rather than weed out the insurgents in the villages.[10] By 1966, SF had begun serving as military advisors to district chiefs and assisting refugees to bolster their defenses against the VC. In 1969, SF soldiers began leaving the sixty camps they had established in South Vietnam. One year later, Creighton Abrams ended MACV's control of the CIDG Program, and a combined council of South Vietnam's Joint General Staff and MACV officials moved montagnard soldiers into border surveillance units commanded by South Vietnam's Special Forces. By January 1971, most U.S. combat troops had departed Vietnam, leaving the South Vietnamese to take control of the remaining U.S. SF camps.

Throughout the SF's tenure in Vietnam, they trained indigenous forces, departing when the village seemed secure enough to leave in the hands of Vietnamese Special Forces and CIDGs. In hopes of drying up the availability of montagnard manpower, labor, and food for the VC, the SF often relocated civilians to villages miles away from their homes. During April and May 1965, SF transferred eight thousand montagnards from forty-eight villages to distant locations with buildings made of bamboo, with metal roofing. Several hundred of these displaced civilians escaped the mass migration, trying to salvage their original villages, though these were likely to fall victim to U.S. and ARVN artillery bombardments. While the relocation was an attempt to secure the safety of the civilians, a reporter on the scene argued that the mass migration was a disruption for the montagnards.[11] Many montagnards actively allied with the SF but, according to Gerald Hickey, they failed to realize that the Americans were simply using them for their own strategic ends.[12]

The many ethnic groups the French originally called montagnards were radically different from the ethnic Vietnamese who dominated the lowland areas. The montagnards physically resembled Cambodians, Malays, and Indonesians, and their languages derived from the Mon Khmer and Austronesian linguistic stocks.[13] The Central Highlanders are composed of twenty-eight different groups, including the Bahnard, Jarai, Katu, Halang, Stieng, and Rhade tribes. Indigenous to the Central Highlands, the montagnards have always dominated the mountainous western border of Vietnam. The ethnic Vietnamese, who arrived in Vietnam later than the montagnards, settled in the lowlands. Throughout the country's history, they were more exposed to colonization and modernization than the highland tribes.[14] As Gerald Hickey reported, historically the montagnards remained "relatively aloof" from the Chinese traditions that played such a prominent role in ethnic Vietnamese society.[15]

The montagnards (known to the U.S. military as "the Yards") and ethnic Vietnamese despised each other, their rift predating the Vietnam War by centuries. Montagnard tribes often settled on land for which they had no legal titles, leading to disputes with the South Vietnamese government. The ethnic Vietnamese viewed the montagnards as uncivilized, lazy savages who did nothing productive on the lands they settled. The montagnards felt threatened by the Vietnamese, whom the ethnic minorities believed were attempting to steal their lands and strip them of their culture. William Colby, who served as head of the CIA and CORDS in Vietnam, compared the ethnic Vietnamese treatment of the montagnards to white Americans' systematic expulsion of Native Americans in the United States; Colby noted that the ethnic Vietnamese pushed the montagnards from the fertile coastal regions to make room for rice harvests.[16] By the late 1950s, the montagnards had long complained that the ethnic Vietnamese–dominated RVN government overtly attempted to keep them out of the country's political discourse. For their part, the ethnic Vietnamese frowned upon the montagnards' refusal to pay taxes. In 1950, ethnic Vietnamese controlled the three provinces with the heaviest concentrations of montagnards. The South Vietnamese government threatened to expel the ethnic Vietnamese province chiefs if they treated the montagnard minorities "too soft."[17]

The montagnard complaints were justified. Ngo Dinh Diem, the leader of South Vietnam from 1955 until his assassination in November 1963, sought to limit the montagnards' involvement in political and military affairs. When the CIA began arming the montagnard minorities in 1961, the South Vietnamese government abhorred the move. The RVN feared that the montagnards would turn their CIA-issued weapons against government officials and ethnic Vietnamese. Responding to pressure from the United States, Diem reluctantly approved the CIA program. In 1964, a group of montagnards created the United Struggle Front for the Oppressed Races (FULRO), an ethno-nationalist revolt that gradually gained popularity among the minorities in the highlands.[18] The leaders of FULRO sought to preserve their civil rights in the RVN government.

On one occasion, montagnard tribes in SF camps rebelled against the RVN, taking sixty South Vietnamese hostages. They were released without harm after the Green Berets negotiated with the montagnards.[19] The ongoing rift between the two ethnic groups was largely out of the SF's hands. They could only hope to contain the uprisings in hopes of preserving the security around their base camps. CAPs never had this problem since they mostly worked with ethnic Vietnamese civilians.

After MACV had diminished the counterinsurgency role of the SF, the U.S. Army did not have any other sizeable force solely dedicated to training local South Vietnamese forces. Yet with the imminent departure of U.S. forces after the 1968 Tet Offensive, MACV and CORDS made improving the effectiveness of territorial forces a top priority. As CORDS began to implement its policies, outside of the program only one U.S. military advisor existed for every 929 soldiers in the territorial forces. From 1967 to 1968, the number of American territorial force advisors surged from 108 to more than 2,000.[20] In April 1968, CORDS and MACV created U.S. Army MATs to enhance the operational effectiveness of the territorial forces. Consisting of five U.S. Army personnel (two officers and three enlisted soldiers), MATs moved from one territorial force unit to another within their target area of operations, usually in one district encompassing numerous villages and tens of thousands of people. Westmoreland had considered expanding the concept of CAPs beyond I Corps to bolster the territorial forces across South Vietnam. Ultimately, however, Westmoreland chose MATs because he believed CAPs would demand too many troops, especially for the large population of the Mekong Delta.[21] MATs required half the amount of troops as CAPs, and CORDS deemed the mobile teams more efficient for Vietnamization since they spent several weeks training one territorial force unit before moving to another. Gen. Nguyen Duc Thang, who in 1968 served as vice chief of staff for the South Vietnamese Joint General Staff, also preferred MATs over CAPs. According to Thang, the PF in CAPs allowed the Americans to do all the work. The territorial forces working with MATs were not as dependent on the U.S. military.[22]

Ultimately, MACV had 354 MATs in South Vietnam. One may use the high number of MATs as evidence that the U.S. Army, MACV, and CORDS spearheaded an organized three-pronged attack on IIIMAF and its use of CAPs. While the program believed the U.S. Army and MACV prevented CAPs from growing, one must consider that CAPs resided exclusively in I Corps. MATs existed in all four corps tactical zones in South Vietnam. During the Vietnamization phase of the war, MACV, CORDS, and South Vietnamese military leaders needed as many troops as possible to train the territorial forces in a relatively short period. Since one CAP demanded twice the manpower of a MAT, the latter offered the most viable and efficient option for manpower allocation. The standard operating procedure of the program did not demand that CAPs spend only a few weeks with a territorial force unit. The Marines remained in the villages until the PF platoons had proven they could operate effectively without the Americans, in some cases for years. During Vietnamization, MACV simply could not afford to have U.S. military personnel spend at least several months with the same territorial force unit. Moreover, placing CAPs in the corps tactical zones south of I Corps would have risked IIIMAF losing its control over the program. The history of the U.S. Marine Corps in the twentieth century shows that its leaders have proven reluctant (at times stubborn) to relinquish control of Marine assets to other military branches. Considering the previous decades of interservice rivalry, it is safe to presume that Marine Corps leaders would have loathed the idea of having elements of their own program placed under the auspices of the army in the southern reaches of South Vietnam.

Throughout its history, the army has always had more troops at its disposal than the Marine Corps, and such was the case in the Vietnam War. In April 1969, when the U.S. military reached its peak strength in the war of 549,000, IIIMAF reported having 78,970 Marines (both enlisted and officers) under its control. That same month, there were 60,419 army soldiers in I Corps alone.[23] By November, the number of U.S. Army soldiers stood at 338,000, compared to 64,400 Marines.[24] To spread CAPs throughout South Vietnam would necessitate either the Marines stretching themselves to the breaking point across the corps tactical zones or allowing army soldiers to man the combined units. With more soldiers and thousands of territorial force units to train, the U.S. Army simply had the manpower advantage that enabled it to implement MATs on such a wide scale. That had nothing to do with interservice rivalry.

Like the CAPs, the MATs trained South Vietnam's territorial forces. The South Vietnamese province chiefs assigned MATs to territorial forces that had been deemed adequate, whereas the program assigned CAP Marines to villages with ineffective PF platoons. In tending to various territorial force units in a district, MATs trained PF that normally operated at the village level, Regional Forces (RF) at the district level, and People's Self-Defense Forces (PSDF), unpaid part-time soldiers operating in their home villages. Created during Vietnamization, the PSDF Program aimed to increase village security by arming local villagers. The territorial forces lived inside each MAT compound and the PSDF lived in adjacent villages. The U.S. Army advisors educated the territorial forces in small-unit tactics and weapons training in addition to providing instruction on ambush and night patrols. Once the territorial unit had proven its effectiveness, the U.S. advisors moved their services to another local force. The average time spent with one unit was thirty to sixty days. Like CAP Marines, MAT soldiers sometimes went on patrols without the territorial forces. The captain in charge of Lt. David Donovan's MAT preferred that the soldiers patrol without the South Vietnamese because he thought the Americans could "move more quietly and operate with more flexibility if we didn't bother with the local units."[25] Donovan explains that tensions between the captain and his subordinates intensified because the commanding officer enforced a policy of patrolling on a daily basis and setting ambush sites every night. Also similar to CAP Marines, U.S. Army advisors offered spontaneous informal training to territorial forces on patrol or in the MAT compound. However, MATs also held formal classes for the territorial forces, whereas the training PF in CAPs received was almost entirely informal. As a CORDS document noted, in CAPs, "the classroom is the 'bush,' and the VC provide the necessary training aids."[26]

The Green Berets encountered military obstacles similar to those existing in CAP and MAT units. Like the PF in CAPs, discipline among the CIDGs came at a premium. However, unlike the PF, the CIDGs received no training whatsoever before the introduction of SF soldiers. Without any previous military instruction, the montagnard forces noisily patrolled through the jungles of the Central Highlands during their initial operations with the SF. Despite their lack of training, however, many CIDGs had a strong dedication to the war effort. Robert Kreger, in 1968 a first lieutenant with an SF A-team, recalls that CIDGs in his unit exhibited loyalty to the United States, which he attributes to the Green Berets before him, who had worked with the same montagnards since the early 1960s. "They were not afraid of anything," Kreger remembers of the

CIDGs. "They were tremendous fighters."[27] Other SF soldiers experienced the worst of the CIDGs. James McLeroy argues that the CIDGs were either worthless, VC sympathizers, or cowards.[28] During his tenure as an officer in an SF A-team, Roger Donlon notes, the Americans' military counterparts seemed largely apathetic, tired of war after decades of being in the middle of conflict.[29]

Payment for CIDGs filtered through the U.S. soldiers rather than the district and village chiefs who controlled CAP PF allotments. However, some SF soldiers practiced the same corrupt policies that plagued the PF payment system. The SF's executive officer, a first lieutenant, managed and distributed money to the CIDGs. Robert Kreger flew to Kontum once a month to pick up a duffel bag full of money to pay his CIDGs. Like village chiefs in control of CAP PF, Kreger provided his superiors with a deceiving roster of "ghost soldiers" to pad the amount of money he received. He pocketed a large portion of the money for his own personal use at whorehouses, dividing the rest among families of CIDG soldiers killed in action.[30]

In addition to the military aspect, the Green Berets had cultural barriers to break through to ensure the success of their units. Like CAPs, SF soldiers had to earn the respect and loyalty of the montagnards by learning and accepting their language and customs. For example, when visitors entered certain montagnard tribal homes, the Americans could not speak until the woman of the house officially greeted the guest.[31] The Green Berets had been taught rudimentary Vietnamese during training at Fort Bragg, but most montagnards did not speak the standard Vietnamese language. The six hundred thousand montagnards of South Vietnam spoke a variety of languages, including Vietnamese, French, Chinese, English, and the dozens of dialects distinct to individual montagnard tribes. Gradually, SF soldiers began to pick up the language. Robert Kreger remembers that he learned the numbers system and sexual-oriented phrases first.[32] SF soldiers adapted well enough to the language over their first two weeks in camp to converse about basic military information.

Many Americans ate indigenous food, while others enjoyed more luxurious, fresh food from U.S. Army base camps. One SF team traded captured enemy weapons at the Fourth Infantry Division base for steaks, chickens, and duck, all airlifted to the camp. As Robert Kreger recalls, "I never ate C rations at all the first year. When we were in camp we ate better than—in camp we ate like kings." Kreger enjoyed baking fresh bread and appreciated being able to store food in freezers and refrigerators, which is why "the mechanic on the generator was such an important man in the camp, too."[33] Kreger's was an atypical experience. Most SF soldiers did not usually enjoy protein-rich poultry from the base camps; they attended village dinners featuring the tribe's local cuisine.

SF medics provided some of the earliest MEDCAPs in Vietnam. Creating a positive rapport with the shamans and sorcerers in montagnard tribes was critical. Montagnard shamans held prominent political roles in their villages. Sorcerers held the power to block the SF medic from tending to wounded villagers with his foreign medical cures. When an SF medic was allowed to perform his medical functions, the local shaman meticulously gazed over his shoulder, burning leaves, beating on drums, and sacrificing animals to help the American succeed. In the montagnard village of Ba To, the locals brought a boy to the resident SF medic hours after he had drowned. When they first found the boy unconscious, the villagers had attempted to revive him by burning his body.[34] If an American medic did successfully treat a sick or injured montagnard villager, he had to acknowledge the assistance of the shaman or sorcerer to allow him to save face.[35] SF medics encountered problems similar to those faced by the corpsmen in CAPs. Like the ethnic Vietnamese in the lowlands where CAPs operated, most montagnards had never seen a physician of any kind. Moreover, SF medics entered villages with high rates of tuberculosis and infant mortality. One of the main differences in the offering of MEDCAPs by SF and CAPs was that the army medic did not hold daily medical calls as the corpsman did. The SF medic roamed among numerous villages and held sick calls once per week.

The spiritual beliefs of the montagnards played a major role in civic action. As one SF soldier told an American war correspondent, "It took us three months before the spirits were convinced that schools were a good thing. But after the spirits were convinced, they built the schools themselves and all we helped out with was a foreman and tin roofing."[36]

SF soldiers often grew fond of the village children. In 1964, the SF group at Nam Dong had their own "mascot," a seven-year-old boy who frequently crawled into the American soldiers' supply room to sleep on a spare mattress. Befriending children had its challenges, considering the

VC's propensity to use them for its own gains. One SF team "adopted" a young village girl, unaware that the VC had threatened reprisals against her family if she did not cooperate with it. As a result, the young girl became "the eyes and ears" of the VC in the village.[37]

David Donovan's *Once a Warrior King* provides a firsthand account of a MAT in the Mekong Delta from the perspective of a lieutenant. Donovan's experiences as a MAT leader in many ways mirror those of CAP Marines and Green Berets. Donovan had to overcome the culture shock that accompanied his arrival to the MAT, dining with villagers, dealing with corrupt South Vietnamese political officials, and participating in civic action. Donovan felt his time in a MAT made him more cognizant of and sensitive to his cultural surroundings in the villages. He came to detest the deplorable animosity that many soldiers outside his MAT displayed toward the Vietnamese. "I have never recovered," Donovan explains, "from the appalling view I got of the conduct of many of my countrymen toward the Vietnamese people." Working closely with the Vietnamese in a MAT, he viewed them as equals: "I had never thought of them as less than ourselves." Donovan had mixed feelings about the effectiveness of the conglomeration of territorial force units working with his MAT. Yet one major difference between CAPs, SF, and MATs that Donovan's book captures regards medical attention. Both CAPs and SF had either U.S. corpsmen or medics available nearby every day. Donovan reveals that in many cases, the team medic was often miles away from the MAT area of operations, tending to other advisory duties. The former advisor recalls that "we all lived with the fear of being wounded and then dying due to the lack of care so readily available to everyone in the American units."[38]

Both the army and Marine Corps developed training regimens to prepare soldiers and Marines for their tours in Vietnamese villages. An examination of their different training systems widens the gulf between SF and MATs, on the one hand, and CAPs on the other. Before entering their assigned CAP villages, the Marines and corpsmen received two weeks of instruction in Da Nang on small-unit military tactics as well as elementary lessons in Vietnamese language and culture. Marine Corps bases in the United States did not offer any schooling specifically tailored for CAP duty. CAP school in Da Nang materialized in an ad hoc manner, as program leaders began to realize the disadvantages of sending nineteen-year-old Marines into a foreign environment without a sense of cultural awareness. On the contrary, Green Berets and MAT advisors completed a much lengthier and detailed training program that began in the United States and continued in Vietnam.

To become a member of the Army SF, a soldier had to first qualify as a paratrooper, which then allowed him to volunteer for the Green Berets. If accepted, SF volunteers endured a lengthy, grueling training process.[39] SF training during the Vietnam era proved far more comprehensive and detailed than the two weeks of CAP school. Depending on the specialty of the Green Beret, training lasted from ten to forty-four weeks.[40]

In the early 1960s the army created its own Special Warfare School at Fort Bragg, North Carolina. The U.S. Army base curriculum focused on unconventional warfare, psychological operations, and counterinsurgency. Each session spanned six weeks, with an additional two-week course specifically geared to officers. Classes such as Senior Officer Counterinsurgency and Special Warfare Orientation provided senior commissioned officers with a general framework on counterinsurgency operations. Officers endured twelve weeks of training in light and heavy weapons, communications, operations, intelligence, and medicine. Special Service officers received one hundred hours of Vietnamese-language instruction, and some handpicked SF soldiers received further training at the Defense Language Institute in Monterey, California. Under the assumption that officers had already undergone a great deal of military instruction before arrival at Fort Bragg, enlisted personnel experienced a lengthier training, especially the medics, who trained for nearly one year. Upon arriving at SF headquarters in Nha Trang in South Vietnam, officers learned of their specific assignments and subsequently attended the Combat Orientation Course, one week of further instruction on the specialties of every member within their unit.

In addition to Green Beret training, the several months of MAT schooling far exceeded CAP school. Before their assignment to MATs, the U.S. advisors underwent twelve weeks of training at the John F. Kennedy School for Special Warfare at Fort Bragg. Every day the advisors in training spent four hours in Vietnamese-language classes, followed by four hours of military instruction. The Special Warfare School also focused on the culture and history of Vietnam as well as historical analyses of the war. Selected advisors extended their language training for another twelve weeks at the Defense Language Institute at Fort Bliss, Texas.[41]

When the advisors landed in South Vietnam, they were assigned to a specific MAT and attended the Di An Advisor School outside Saigon for four weeks, receiving more than a hundred hours of training in weapons and tactics, including thirty-six hours dedicated to Vietnamese-language instruction.[42] South Vietnamese military chaplains lectured on the basic geography and history of Vietnam, including as well a brief description of Vietnamese religious beliefs.[43] Army advisors received four hours of language instruction every day in addition to introductory training on the World War II–era weapons of the local forces. First Lt. Ron Milam remembers "having a heck of a time trying to learn how to field strip an M1 Garand because it just wasn't natural for me."[44] The school hoped that the brief language instruction would allow the American advisors to engage in simple conversations with the Vietnamese, providing a solid foundation on which to build their linguistic skills.

In 1968, Gen. Robert Cushman, as IIIMAF commander, instituted mobile training teams (MTTs) in I Corps, which tactically were mirror images of army MATs. The use of MTTs coincided with the development of the Combined Unit Pacification Program that employed Marines from infantry units with CAP principles to train PF platoons. With the adoption of MATs in April 1968, the Marines used squads of CAP Marines to train non-CAP PF in I Corps for a two-week period. After giving instruction on infantry tactics over those two weeks, each MTT then moved to the next non-CAP PF platoon in its area of operations. By October 1968, eight MTTs totaling ninety-eight Marines and corpsmen had trained forty-five non-CAP PF platoons in I Corps.[45] Although the number of MTTs in I Corps remained miniscule compared to the number of CAPs, Victor Krulak deemed the entire MAT concept "worthless."[46] MTTs did not last long, as within the next year army MATs began replacing them. In his monthly FMFPAC reports, Krulak never criticized the Marines' version of the army MATs, but he did argue that in just two weeks, the PF associated with MTTs did not receive the same practical experience as the local forces in CAPs.[47]

The Marines' use of MTTs may have mirrored army MATs in some respects, but when one compares the training process of the Green Berets and MACV advisors to the schooling for Marines and corpsmen in the program, they stand at polar opposite ends. The Marine Corps did not have a military base in the United States such as the army's Fort Bragg to prepare trainees for counterinsurgency warfare. The enlisted Marines and NCOs who would enter the program were trained foremost as riflemen. They had not received any extensive language training, nor did they attend formal classes introducing the intricacies of Vietnamese culture. Compared to the program, the U.S. Army had a much more extensive training process for the officers and enlisted soldiers alike who would serve in either the Green Berets or mobile advisory teams.

As the program grew during the first two years of the war, its leaders began to realize the need to screen and train the Marines who had volunteered for CAP duty. The program did not want to risk the political and military backlash that might follow a CAP Marine's insensitivity to villagers and their customs. The selection and training criteria the program ultimately formulated in an ad hoc manner contained numerous inadequacies, but nothing could have fully prepared the Marines and corpsmen for the assignment that lay ahead of them.

CHAPTER THREE

Becoming a Combined Action Platoon Marine

It has been clear from the beginning of the Combined Action Program that special training is necessary for the men to be placed in the villages, for, despite a long tradition of Marine involvement in constabulary type operations, knowledge gained from this experience has had no effect on basic recruit indoctrination and conditioning, which is still aimed at the more classic Marine role of conventional aggressive operations.

—Bruce C. Allnutt

Until 1967, Marines entering CAPs did not have any formal schooling in Vietnam on the distinct military and cultural environment they would encounter in the villages. Early in the war, Marine commanders at IIIMAF and FMFPAC headquarters had become concerned about the rising number of physical altercations between American GIs and South Vietnamese civilians in I Corps. Facing a versatile and elusive enemy often disguised as civilians, American GIs began to perceive all South Vietnamese as potentially allied with the enemy. Combined with an ignorance of Vietnamese culture, American GIs' general disdain for the South Vietnamese population confronted program leaders with a demanding yet necessary task: groundwork for the establishment of amicable U.S.-civilian relations at the village level was critical to include in the selection and training of Marines for CAP duty. The criteria for becoming a CAP Marine changed several times during the course of the program's implementation (1965 to 1971). In an ad hoc, trial-and-error manner, the program's Marine leadership adjusted the criteria for selection according to manpower availability in I Corps. Until 1969, the program plucked Marines from units already in Vietnam. Although the Marine Corps recognized the usefulness of pacification and counterinsurgency, units not dedicated to "the other war" still needed manpower to fend off the larger enemy battalions in the I Corps countryside. The major quandary for IIIMAF and the program was devising a way to increase the number of CAPs while keeping the infantry and support units adequately stocked with Marines. This was a problem in part because units that spared Marines for CAP duty did not automatically receive replacements from newly trained recruits. With program quotas to satisfy and a desire to keep their best Marines, commanding officers often chose personnel whom they wanted to discard rather than those who fit the program's criteria. Not until 1969 did the program switch from choosing Marines in Vietnam to assigning them in the United States before they became acclimated with non-CAP units and their respective commanding officers.

In 1967, to help alleviate the likely culture shock for the CAP Marines about to enter their assigned villages, the program designed a mandatory two-week school with a curriculum focused on small-unit military tactics and rudimentary training in Vietnamese language and customs. Considering the relatively short period of CAP school, its graduates were mere neophytes with regard to Vietnamese culture and language. For Marines entering the program from units in Vietnam where criticism, negativity, and racism toward the Vietnamese people were rampant, cultural training was necessary, but as a colonel in the program acknowledged, "It was not possible to transform these Marines into linguists or cultural anthropologists overnight."[1] During the first three years (1965 to 1968) of the war, when both the VC and NVA gradually increased their infiltration into the countryside of I Corps, IIIMAF needed Marines in the villages quickly.

Duty in a CAP presented cultural and military obstacles that warranted a myriad of adjustments by the Americans, who had to adapt to life without electricity, running water, sanitation, an abundance of Western medicine, or persons who spoke their language. Militarily, performing counterinsurgency and pacification tasks contrasted sharply with the infantryman's routine search and destroy missions that infamously characterized America's war of attrition in Vietnam.[2] Although American line units spent days in the jungles of South Vietnam away from their base camps, they never lived in the villages. Nor did they interact with South Vietnamese villagers on a daily basis, much less try to understand their culture. Said one Marine corporal, "Training back in the States leaves out one big thing—the Vietnamese."[3] Although of dire importance for the program's success, the selection process and CAP school played only minor roles in helping the Marines transition to life in the villages. Actually living in a CAP village and socializing with the locals provided the best training.

Twentieth-century U.S. military history is filled with instances of Americans, both civilian and military, dehumanizing their enemies, characterizing them with racist qualities and animal-like attributes. John Dower's *War without Mercy* details this process in the Pacific theater during World War II. In what Dower calls a war of extermination, Americans depicted the Japanese as apes and monkeys, using racial slang such as "Nip" and "Jap" to describe the people of Japan. According to Dower, the dehumanizing of the Japanese and the acceptance of their subhuman characteristics softened the killing process for the American military.[4] Gerald Linderman has classified the Pacific theater of World War II as "War Unrestrained." Linderman argues that the Japanese refusal to surrender and their desperate suicide attacks on air, land, and sea confirmed American beliefs that the Japanese were not human.[5] In his extraordinary memoir about his time in the Pacific theater, former Marine E. B. Sledge wrote that the "attitudes held toward the Japanese by noncombatants or even sailors or airmen often did not reflect the deep personal resentment felt by Marine infantrymen."[6]

In the Vietnam War, "gook" became the popular American racial slang for the Vietnamese. The derogatory term did not originate in Vietnam. David Roediger traces its use back to the U.S. Army's encounters with Filipinos during the Spanish-American War. From the Philippines to Haiti, World War II, and the Korean War, the U.S. military used the term *gook* to describe both friendly natives and the enemy. By September 1950, just four months after the start of the Korean War, "gook" had become so commonly used by American troops that Gen. Douglas MacArthur, the commander of all United Nations forces, ordered the discontinuation of the term out of fear that the racist language would devalue American democratic ideals as perceived by the South Koreans. However, MacArthur's command apparently fell mostly on deaf ears, as Americans continued to use the term liberally for the remainder of the war. Although the term had been a part of the U.S. military's vocabulary for decades, according to Roediger, the Vietnam War solidified it as a modern reference to Asians.[7]

As Mark Philip Bradley has demonstrated, American perceptions of the Vietnamese people as inferior reach back into the French colonial years after World War I. Descriptions of Vietnam and its people mirrored the general American perception of all Asian civilizations. American observers deemed the Vietnamese lazy and effeminate, with the men lacking military prowess. One American, noting the living conditions within the villages of Annam, the region that encompassed the central part of French-controlled Vietnam, blared, "Annamites at best are never clean, but sickness shows up this trait in its most revolting form."[8] In the 1930s, American observers began to criticize the French for failing to improve the lives of their colonial subjects, whom the Americans deemed unfit for self-improvement due to their racial characteristics and physical environment.[9]

The dehumanizing of America's enemy using racist rhetoric in the Vietnam War began when recruits first entered boot camp, where drill instructors constantly referred to the Vietnamese as "gooks" and "slant eyes."[10] Peter Kindsvatter argues that the dehumanizing element of boot camp was different for World War II recruits than it was for those going to Vietnam. Due to the widely publicized atrocities committed by the Japanese and Germans, World War II GIs had entered boot camp with a preconceived belief that their enemy was evil. Yet Vietnam recruits entered boot camp with relatively less animosity toward an enemy and a country they knew nothing about. Kindsvatter concludes that drill instructors used "gook" to rile aggression and hatred among the recruits toward the enemy.[11] Fresh American arrivals to Vietnam were ordered to "kill, kill, kill the gooks!"[12] Even before experiencing the frustration and exhaustion of war, many American GIs arrived in Southeast Asia believing that "the only good Vietnamese is a dead Vietnamese."[13] Michael Peterson, who ultimately joined the program, wondered how Marines could accept the Vietnamese peasants as human beings when "we were fed 'Luke the Gook' from Boot Camp onward."[14]

After boot camp, Marines attending advanced infantry training (AIT) practiced fire maneuvers, performed intense physical training, and sharpened their marksmanship skills. As part of the training, the Marines walked through a simulated Vietnamese village, replete with pop-up targets and fake booby traps hidden among the bamboo huts. Thus, the training portrayed Vietnamese villages as dangerous places that the Marines should approach with caution and suspicion. In retrospect, Vietnam veteran Tom Esslinger wishes he had received better training regarding Vietnamese villagers. "I would have had a much better understanding of how to treat a village," Esslinger reveals, "if I would have understood more about their religion, their culture, their veneration for old people, their attachment to the land."[15]

From 1964 to 1968, U.S. Army general William Westmoreland served as the commander of MACV, affording him operational control of all U.S. forces in South Vietnam. Westmoreland and his fellow army colleagues at MACV headquarters in Saigon implemented a war of attrition. In this strategy, Westmoreland's and U.S. Secretary of Defense Robert McNamara's measuring stick for success was attaining a high, favorable kill ratio in which the Americans were to kill ten of the enemy for every U.S. death. Employing superior technology and massive firepower, Westmoreland ordered the U.S. military to search for and destroy large enemy units, the purpose being to achieve a high enemy body count. Westmoreland sent the infantry into the jungles of Vietnam for no other purpose than to find and kill large enemy units on sight. The conventional wars of America's past had proven that annihilating the enemy's army would bring success for the United States. Thus, commanders wanted to relay successful "after-action" reports to their superiors—and dead Vietnamese bodies equaled success. For Westmoreland and McNamara, as long as the U.S. military actively engaged the VC and NVA in battle, the combined American–South Vietnamese military effort would emerge victorious. Yet as the war progressed under Westmoreland's command, American air, land, and naval forces became frustrated that their adherence to the war of attrition strategy was not forcing the enemy to capitulate.

Countless American units on search and destroy missions marched for days and in some cases weeks without making any contact with the enemy. Having a greater knowledge of the climate and terrain of Vietnam, the enemy generally chose when and where a firefight took place. Research conducted during and after the war found that the enemy initiated from 65 percent to 75 percent of all contacts.[16] The NVA and VC willingly surrendered land to American forces during battle in exchange for U.S. casualties. The North Vietnamese surmised that war in the jungles of Vietnam would result in a high U.S. death toll, eventually forcing an outraged American populace to call for an end to the conflict.

The frequent inability to initiate contact with an elusive enemy proved mentally maddening and physically exhausting for American grunts. Describing the frustrating military environment for American GIs in the field, Marine first lieutenant Philip Caputo has written, "The guerrillas were everywhere, which is another way of saying they were nowhere." As the junior officer in charge of a Marine infantry platoon, Caputo participated in numerous search and destroy missions. To ensure the success of his unit by reporting high enemy body counts, Caputo's superior officer informed him of the proper protocol for counting dead bodies: "If he's dead and Vietnamese, he's VC."[17] Disregard for possible civilians amid a cluster of "enemy" bodies on search and destroy missions occurred frequently. Describing the aggravation of infantry duty, Jerry Goller concludes that when the unit finally did make contact with the enemy, the Marines would "take out their frustration of the previous few weeks." Goller emphasizes, "It's not good for people to have their only goal to be to see how many people they can kill that day."[18]

During search and destroy missions, grunt units were bound to encounter at least one of the nearly three thousand villages in South Vietnam. Infantry units often entered a friendly village environment but departed under fire. In 1968 Marine lieutenant Lewis Puller Jr., the son of the "Marine's Marine" Lewis "Chesty" Puller, led his infantry platoon into what seemed like a friendly village. The Marines handed out candy and cigarettes and ate food offered by the villagers while the corpsmen tended to villagers' wounds. Just as Puller's unit was leaving, the Marines came under small-arms fire from within the village. None of the Marines was wounded, but the lieutenant had to convince his angry unit not to retaliate. "It was hard to believe," Puller remembered, "that the same villagers who had just shared their meal with us would now allow us to be used for target practice." American ignorance of village life sometimes caused offense to the Vietnamese. Passing through a village, a member of Puller's platoon unzipped his pants and relieved himself in a small hole, unexpectedly earning the scorn of the villagers; he had unwittingly contaminated the village's water supply.[19]

The key to success for the VC rested among the rural population. The VC constantly entered villages to collect taxes, stash supplies, politically indoctrinate villagers, recruit civilians, and hide between missions. Grunts, therefore, had the opportunity to find enemies hidden among a village's civilians—and knowing VC tactics, they usually assumed the worst. With "search and destroy" their primary objective, "humping" in the jungle for days at a time, with at best four hours of sleep per night, combined with the belief that "if it's dead and Vietnamese, it's VC," some irritated grunt units showed little sympathy toward village civilians suspected of supporting the enemy.

The grunts for the most part had little or no desire to win "hearts and minds." Col. David Hackworth, who in 1969 took command of an army battalion in the Ninth Division, grew annoyed at the army's "tired commercial" of winning hearts and minds. Hackworth believed that for the grunts, "the best way to deal with the Vietnamese was to grab them by their balls. Then, they knew, their 'Hearts and Minds' would follow."[20] A 1966 report on American attitudes in I Corps toward the Vietnamese revealed that Marines rarely thought about the negative impact their actions could have on civilians' perception of the American military.[21] Peasants from the village of My Thuy Phuong, southwest of Hue, remember American GIs hitting their children in the head with C rations and shooting water buffalo.[22] The consequences of these incidents and others that were reported—burning homes, slapping elderly people, and stealing religious items from homes—rarely crossed the grunts' minds.[23]

Caputo's platoon burned a hamlet to the ground upon learning that the villagers had aided the VC. Caputo recalls that his Marines were learning to hate. Later in his tour, Caputo and his platoon returned to the hamlet they had burned. The junior officer was astonished by the villagers' seeming indifference as the Marines entered the hamlet for a second time. Caputo's initial regret for what his Marines had done turned to contempt. "Why feel compassion for people who seem to feel nothing for themselves?"[24] As a grunt's tour in Vietnam lengthened, he often became desensitized by the war, and the erosion of his attitude and self-control soon followed.[25]

The thoughts of Marines who entered the program after serving in infantry units mirrored the typical experiences of grunts. Before joining a CAP in 1967, Igor Bobrowsky, an infantry squad leader, had witnessed Marines burning houses and shooting Vietnamese civilians. During his tenure as a grunt, Vietnamese civilians meant nothing to him: "If it was hostile it didn't matter if it was a coconut or a human being or a buffalo or a tree, it was all the same. I really didn't think of them as anything. They were just there."[26] Barry Goodson had a similar experience, entering CAP school from an infantry unit. When Goodson realized one of his CAP school instructors, a South Vietnamese officer, was intelligent, he thought to himself, "It's about time. Up until now we only thought of the people as simple idiots, or animals we could slaughter without a second thought."[27]

By 1966, the Marine Corps had grown disturbed by physical encounters between Marines and South Vietnamese civilians. It was obvious even to some non-Marine observers that Lt. Gen. Victor Krulak wanted to improve the American GIs' understanding of the Vietnamese people in I Corps.[28] Krulak was concerned that the Americans' lack of understanding and empathy would have a negative impact on South Vietnamese civilians' support of U.S. forces.

When in 1966 Krulak crossed paths with U.S. Navy chaplain Lt. Cdr. Richard McGonigal, who headed a study that promoted better understanding between Marines and the Vietnamese, the FMFPAC commander quickly pursued ways to broaden the scope of the project. The project had begun in 1965 as a research venture for chaplains to illuminate the religious reasons behind the behavior of the Vietnamese. More important for Krulak, the project's chaplains wanted to relay their findings to Marines in I Corps in hopes of reducing the cultural misunderstandings that had led to unfortunate incidents in the villages.[29]

By 1966, the chaplains' religion-based research had turned into a more comprehensive project encompassing a wider range of cultural elements in Vietnam, with McGonigal serving as the main advisor for the newly named Personal Response Project. McGonigal, the only American to visit all 114 CAPs in Vietnam, compiled several surveys in I Corps to ascertain the attitudes of both Americans and South Vietnamese. With the wholehearted support of Krulak and Walt, the chaplains of the Personal Response Project hoped that the information gathered from their findings would influence Marines to "act with understanding, concern and responsibility in relationships with indigenous citizens."[30] Winning "hearts and minds" was not enough for the creators of the Personal Response Project. The chaplains wanted the Americans to view the Vietnamese as equals.[31]

By August 1966, McGonigal had held numerous conversations with

Marines and corpsmen and had concluded there was an urgent need to increase awareness of the Personal Response Project. ARVN officers, PF, and civilians had already begun to react negatively to the apparent American contempt for the South Vietnamese. McGonigal noted, "Expressions of distrust and disdain far outweigh expressions of understanding and tolerance."[32]

To market the usefulness of the project, chaplains presented their findings at all levels of IIIMAF, from Walt's headquarters to bunkers in the field. Lecture subjects centered on Vietnamese religions, civic action, and personal conduct, along with explanations of the purpose and objectives of the Personal Response Project. In addition, the Personal Response Project spread the word via monthly newsletters. These publications updated Americans in I Corps on recent successful interactions between Marines and South Vietnamese. Moreover, the newsletter attempted to explain why the South Vietnamese behaved as they did. For example, the October 1966 issue featured a section detailing why the Vietnamese did not always thank the Americans for friendly services.[33]

By August 1966, 50 percent of the GIs in I Corps had heard lectures from some of the multitude of chaplains assigned to the project.[34] But in the fall, the Personal Response Project was still encountering complacency from Marine commanders who were reluctant to expose their troops to the chaplains' lectures. Despite an FMFPAC order in March to allow the lectures, few commanders complied. In October, McGonigal reported that the general in charge of operations and training for the First Marine Division had informed his Marines that the unit's chaplain was prepared to deliver a lecture on religions in Vietnam.[35] However, there was no requirement to attend.

According to a Personal Response Project report in 1967, McGonigal had just cause for concern regarding the Marines' feelings about the Vietnamese. The report surveyed five hundred Marines, many coming from nine infantry battalions and nine noninfantry battalions in Vietnam. The results showed that only 37 percent of the Marines reported "liking" the ARVN and PF and 43 percent "liked" Vietnamese civilians. Perhaps the most alarming statistic from the report was that 66 percent of the sample had "violent dislikes or mixed feelings" toward the South Vietnamese.[36] Despite these disheartening statistics, McGonigal refused to give up on the expansion of the project. By 1967, the project had created councils at each division's headquarters and jump-started a Personal Response Project school for the Third Marine Division.

Although Krulak supported the project, others at FMFPAC headquarters deemed chaplains unfit for the duties outlined by the Personal Response Project. By 1967, when the mission of the Personal Response Project changed from a purely religion-related venture to one that incorporated attitude and behavioral analyses, dissenters, including Marine officers and Navy chaplains, wanted to delegate the jobs to U.S. State Department officials. They believed that chaplains should serve a strictly religious function; examining Marines' general attitudes was irrelevant to their primary jobs as spiritual guides. Yet McGonigal continued to champion the chaplains' roles in the Personal Response Project, asserting that in Vietnam "lives are being lost by this communication of poor attitudes and I think that concerns for the project's reputation or the project officer's future pale in comparison with the immediate and urgent importance of attention in the field."[37]

CAPs presented a unique glimpse into a world in which Marine survival hinged on creating a friendly rapport with civilians. CAP school welcomed the Personal Response Project chaplains, who frequently lectured, organized village simulations with CAP student role-players, and made final presentations at graduation. Both students and teachers at CAP school saw great value in the Personal Response Project. One of the chaplains at CAP school remembered "excellent class participation and lively discussions."[38] During the month of March 1969, personal response teams allotted fifteen days to CAP school, four to IIIMAF headquarters, and one to the twenty-ninth civil affairs unit.[39]

From October 1966 until June 1967, the Personal Response Project conducted research on the attitudes of Marines at the NCO leadership school in Okinawa. Part of the report compared the attitudes of the NCO students with the consensus of opinions provided by non-NCO Marines in I Corps and Marines in CAPs. When asked if they liked the Vietnamese, CAP Marines scored in the highest percentage regarding both the South Vietnamese military and civilians.[40] The CAP Marines from this particular study had never attended the formal CAP school, formally created in 1967. Thus, it seems that their time in the CAP villages, much of it spent interacting with the people, had a significant impact on their transformative perceptions of the Vietnamese. Ron Schaedel entered a

CAP in 1966, never having received formal cultural training in CAP school. Nonetheless, he recollects gaining a better respect for the Vietnamese people as his time in the village progressed.[41]

For two months, beginning in December 1966, Personal Response chaplains visited forty-eight CAPs, with special attention given to the personal relationships developed in the villages. Surveys encompassed both Marines and PF. The officer for the report also visited hamlets without CAPs, noting vast differences between the two. In CAP hamlets he found a much cleaner and more sanitary environment, with the Americans upholding Vietnamese traditions. In the hamlets without CAPs, the project officer observed, it was "business as usual." The findings concluded that "the cutting edge of the whole pacification effort, as we see it, is the combined action unit. It is with the combined action unit that we are most concerned in establishing trust and mutual respect."[42]

By the time William Corson had ascended to the program's director position in 1967, he had maintained frequent dialogue with McGonigal. Corson was instrumental in ensuring that McGonigal and the Personal Response Project played integral roles in educating CAP school attendees and improving the program as a whole. McGonigal's perceptive analyses of CAPs, which can be attributed in part to the chaplain's cordial relationship with program leaders, helped to lay the groundwork for the official CAP selection process that Corson ultimately adopted. Personal Response Project reports had detailed particular characteristics of Marines that had a tendency to create and sustain favorable attitudes with the Vietnamese. McGonigal concluded that persons with more than sixteen years of education tended to be more critical of the Vietnamese than those with only a high school diploma. In an interview, McGonigal argued that educated professionals such as doctors were "used to taking care of people in a nice, clean clinic. People come by appointment. And here he walks into his MEDCAP where there's 50 women throwing their babies in front of [him]; they haven't used the soap they were given two weeks ago—sold the soap for food. The doctor can't handle that and the dentist can't either. They treated the people quite contemptuously."[43] The program never specifically targeted uneducated Marines for CAP duty, but many of the teenaged enlisted personnel in CAPs had never graduated from high school.[44]

The overall impact of the Personal Response Project on Marines in the field is difficult to assess. The official history of the U.S. Marines in Vietnam shows that both Marine divisions in 1968 reported that the Personal Response Project had reduced the number of nonoperational incidents with Vietnamese civilians.[45] However, it is virtually impossible to determine if the project was exclusively responsible for preventing individual Marines from committing violent acts of aggression against nonhostile civilians. Still, the Marines at IIIMAF headquarters, and later in the war MACV officials, saw great value in the project. The Personal Response Project never materialized in the corps tactical zones south of I Corps to the level it reached under the guidance of IIIMAF. In late 1969, as the U.S. military began its withdrawal from South Vietnam, MACV commander Gen. Creighton Abrams voiced his regrets for failing to use the resources of the Personal Response Project. "I kind of wish we'd all been on that, you know, back at the beginning. . . . A fellow that can't respect the Vietnamese has got to get into some kind of job where they won't see him."[46]

The roots of the program's selection and training procedure stemmed from Lt. Paul Ek, who in August 1965 took command of the first CAP near Phu Bai. Ek, who had previously served in Vietnam as a Marine advisor to U.S. Army Special Forces, proved a worthy leader for the first CAP. Before taking command, he had attended a two-month course on the Vietnamese language; it was not long enough to attain fluency, but he could get his point across to a civilian.[47] In creating the first CAP, Ek pursued highly motivated volunteers who had prior experience working with the Vietnamese people, and subsequently put his men through a two-week course on their military roles in working with the PF as well as training in Vietnamese language and culture. The culture education was intensive, investigating distinct rules and customs, which varied from one village to another.[48] The CAP commander feared that American ignorance of acceptable behavior in the villages could translate to disaster for the Marines. He hoped the training would allow his Marines to "live with [the villagers] in a close relationship, not as an occupational force, but as members of that village."[49]

Within a year of the establishment of Ek's unit, the program concentrated on creating eligibility requirements for Marines seeking to join a CAP. By the end of 1966, the program had grown to fifty-eight CAPs, a drastic increase from the six platoons in existence just one year earlier.

This massive growth forced program leaders to solidify qualifications for Marines entering CAPs. On a cultural level, plucking random Marines from the field without any sort of background check could prove disastrous, perhaps leading to apathetic personnel with disciplinary problems. The first requirement for finding CAP Marines called for commanding officers in the field to find "mature and highly motivated" volunteers within their units. Yet, realizing that definitions of these qualities might vary from one commanding officer to another, the program further specified the criteria. In 1967, Corson created a list of eleven criteria, among them that a Marine must: have been in country for at least two months or have served on a previous tour; have a minimum of six months remaining on his current tour or agree to extend to meet this requirement; be a volunteer; have received no nonjudicial punishment within the past three months; be recommended by his commanding officer; and have not received more than one Purple Heart on his current tour. Persons who failed to meet all the criteria yet came recommended by their commanding officers had to pass an interview with a committee of program officers, the intent of which was to gauge the Marine's personal desire to serve in a CAP and work with the South Vietnamese.[50] The 1967 criteria did not require any prior combat experience in line units, as it had during the first two years.

The selection and screening process contained loopholes. When Col. Edward Danowitz became director of the program in October 1968, he established quotas for both Marine divisions in I Corps. Danowitz called for as many as twenty-five volunteers from each division during certain months. Yet the divisions did not receive immediate replacements for those who volunteered. Thus, it came as no surprise that with quotas to satisfy, commanding officers filled CAP ranks with unqualified persons. Rarely would a commanding officer willingly part ways with his best Marines. He had to consider the detrimental impact on his own unit. A commander actively seeking to send much-needed bodies from his unit into the program could jeopardize his own unit's safety. In short, Marine infantry commanders viewed their orders to fill CAP rosters as a necessary burden they had to tolerate.[51] To avoid damaging the military capability and cohesiveness of their units, commanding officers often "volunteered" unit misfits for CAP duty. Cases emerged in which sergeants of line units submitted lists of "deadwood" Marines to their commanding officers for transfers to CAPs. During his company's formation, Tom Harvey from the Ninth Engineer Battalion volunteered for the program but was bypassed the first time because one of his unit's "shitbirds" was chosen instead by the staff NCOs.[52] The program director's staff monitored the records of the chosen Marines before they attended CAP school, but unrecorded disciplinary problems in line units escaped that screening process. Moreover, the program failed to create a system for replacing Marines in CAP villages who repeatedly committed acts detrimental to the unit and civilians. CAP commanders had the authority to rid their CAPs of ill-disciplined and culturally insensitive members, but the program tended to shuffle these counterproductive Marines from one village to another.[53]

Some commanders opted for Marines who seemed indifferent to the assignment. Col. J. E. Jerue, commander of the Ninth Marine regiment during the war, admitted that Marines from his unit volunteered for CAPs simply by not objecting.[54] Other grunt commanders falsely informed their Marines that CAP duty was only a temporary assignment. Igor Bobrowsky entered the program from an infantry unit after his commanding officer assured him that he would return to his original outfit before the end of his tour. Bobrowsky had been in his CAP for weeks before he realized his time in the program was permanent.[55]

In 1969, the administration of the CAP participant selection process moved from Vietnam to the United States. Two years before, IIIMAF's focus in the war had turned to northern I Corps, where the NVA had increased its attacks. As a result, the growth of the program stalled momentarily, as manpower and resources were committed to infantry, artillery, and support units fending off the NVA near the demilitarized zone.[56] However, in 1969, U.S. manpower in South Vietnam reached its peak at more than five hundred thousand personnel. During the apex of U.S. involvement, the program selected CAP personnel in the United States before arrival in South Vietnam. At Camp Pendleton in California, program officials screened the records of the selected Marines, personally interviewing all prospective CAP members. The downside of this placement system was the inexperience of these Marines in battle conditions: if they froze in combat, it would prove disastrous in a small unit like a CAP. A Marine Corps study dissecting the 1969 selection process found that combat-hardened Marines saw no problem with sending inexperienced Marines into CAP duty, arguing that every battle is new to everyone involved.[57] Placing Marines with little combat experience in CAPs was nothing new for the program. Beginning in 1967, the program welcomed volunteers from noninfantry units, where combat was a rarity. Judging from the generally upbeat CAG command chronologies and FMFPAC's glowing reports of the development and success of CAPs during the first four years, pumping "green" Marines into the villages was not a major concern for the generals and colonels managing the program.[58]

The thought of routinely having a cot and hot chow in the villages ignited the belief in infantry units that assignment in a CAP was "soft duty." Infantry commanders, who had a generally negative perception of those serving in the program, exacerbated the misperception that CAP Marines had a relatively easy life in the villages. As the executive officer of the program and commander of Second CAG in 1970, J. J. Tolnay, points out, Marine infantry commanders described CAP Marines as "a group of undisciplined and obviously un-Marine like group of hippies who had gone native."[59] Numerous volunteers for CAP duty did not have a personal commitment to helping the Vietnamese; they were searching for relief from their prior assignment in Vietnam, which in many cases was the infantry. Many grunts welcomed the prospect of ditching the grind of the infantryman's daily routine in the Vietnam War for what seemed to be a laid-back atmosphere in CAP villages. Grunts who wanted to leave their infantry unit but failed to meet the CAP criteria often passed the subsequent interview process by knowing beforehand the answers the board wanted to hear. Marines simply had to convince the interviewees of a personal desire to help the Vietnamese people.[60]

Many Marines entering the program, whether voluntarily or not, had little knowledge of what assignment to a CAP entailed. CAPs rarely crossed paths with Marine and army infantry units on the move. When Barry Goodson arrived at CAP headquarters fresh from a grunt unit, he noticed a sign on the door: COMBINED ACTION PLATOON ORIENTATION SESSION. Goodson's response? "So that's what CAP stands for."[61] Some of the infantry had heard of CAPs but did not concern themselves with details. Even infantry units assigned as a reactionary force for CAPs in their area of operations did not fully understand the purpose of the combined units.[62] This is not to argue that all Marines in Vietnam lacked an understanding of the hardships encountered in CAPs. Marine grunt David Crawley rejected the program's call for volunteers on the grounds that "I kind of like having hundreds of Marines around me."[63] Jerry Goller volunteered for the program after serving in the infantry for one year because "after spending a year running around trying to find people to shoot, I wanted to, see, do something positive to alleviate some of the misery in that country."[64]

Rather than transferring Marines straight from line units to the villages, the program deemed it necessary first to educate the Marines and corpsmen on the military and cultural norms they could expect in CAPs. Beginning in 1967, all Americans who would ultimately land in CAP villages first attended a two-week CAP school on "China Beach" in Da Nang. Both American and South Vietnamese instructors taught classes on small-unit tactics, weapons used in the villages, and the Vietnamese language and culture.[65]

CAP school curriculum frequently changed, depending on the military occupational specialty of the Marines in the classroom. During particular periods when a high volume of noninfantry personnel volunteered, the curriculum favored the military aspects of CAP duty under the assumption that the students were not as combat hardened as the grunts. In 1969, when Marines entered the program in the United States without any combat experience, roughly 60 percent of the curriculum focused on military subjects.[66] For the duration of the program, regardless of the CAP students' combat experience, Marines at school always received lessons on the various weapons used in a CAP and the proper procedure for calling in air and artillery strikes as well as medical evacuation (MEDEVAC) requests. In addition, CAP students received refresher courses in map and compass reading.

The cultural customs lessons detailed peculiar physical movements customary to Americans but deemed unacceptable by the Vietnamese. For example, villagers frowned upon persons patting children on the head because the action held a sacred meaning. Instructors also taught Marines to avoid pointing their feet directly at villagers since it was considered an insult by the Vietnamese.[67] Barry Goodson remembers learning that villagers would steal from nearby villages, strangers, or neighbors, and that the Vietnamese judged people favorably according to their thievery abilities. This was difficult for Americans to understand and accept.[68]

The diet of Americans in the field consisted of instant coffee, cocoa, and C rations, which featured entrees of American cuisine, including the infamous ham and lima beans. If the C rations supply had been exhausted during a mission, a unit had to resort to killing animals, or in some cases insects, to supply fuel for the next day's march. The Americans entering CAP school, used to devouring C rations without any care as to whom they might be offending, learned the intricacies of proper dinner etiquette in the villages. "Everybody knows how to eat," Corson declared in an interview after the war, "but you didn't know how to eat a meal in a Vietnamese family."[69] When invited by a Vietnamese family for dinner, a frequent occurrence in CAPs, the Marines learned they must always leave food on their plate. The Vietnamese perceived an empty dinner plate as a failure to satisfy the hunger of their guests.[70]

Before becoming director of the program in 1967, William Corson had commanded a Marine tank battalion with the primary goal of pacifying a village in the province of Quang Nam. Corson noted that the success of his unit in the mission was due in large part to a game called elephant chess (co tuong), similar to checkers. The game became so popular among both the Americans and South Vietnamese men in the village that Corson organized an elephant chess tournament. He found that infusing Americans into the game heightened their stature and acceptance in the village. As the program director, he introduced CAP Marines to elephant chess. For the Vietnamese, elephant chess was a status symbol that reflected a man's cultural prowess. Corson hoped that playing the game would elevate the Marines' status in the villages.[71]

Marines received only the bare essentials of language training. Having a firm grasp on the language would have assisted in creating an amicable rapport with the civilians and PF much sooner. Yet by July 1969, the school committed only 13 out of a total of 123 hours to Vietnamese-language training. Americans entering their CAP villages often witnessed program veterans joking and laughing with the Vietnamese. Yet as Bruce Allnutt argues, "An observer who speaks Vietnamese himself, however, will soon begin to notice that the vocabulary used in such interchanges is extremely limited, and that the content of such communications is restricted to very ordinary phrases common to the environment that are used repetitively." Further complicating the Marines' language deficiencies were the differences between pronunciations of Vietnamese in different parts of I Corps. The regional idiom of I Corps is distinctly different from that of Saigon, the one most widely taught at CAP school.[72]

The structure of language classes made learning difficult. Classes crammed with more than one hundred students, in which the instructors had to shout to be heard, made the educational experience far from ideal. Instructors, comprised of both American and South Vietnamese military officers, also had to squeeze an inordinate amount of linguistic information into a short period. The hope was that if the teachers provided the basics of the language, such as the tones, spelling, pronunciation, and syntax, the Americans could learn more efficiently in the villages. For example, students would repeat words and phrases in Vietnamese back to the instructor, much as in foreign-language classes in American primary schools. Graduates of CAP school left with a language efficiency that one CAP Marine called "jive Vietnamese."[73] Corpsman John Nicols noticed the language dilemma during one of his routine MEDCAPs: "I can't ask which side does it hurt on, is there a sharp pain on that side or about vomiting and diarrhea."[74]

Marines who never joined CAPs could qualify for a twelve-week Vietnamese-language course if they scored high marks on language-proficiency tests given during boot camp. In Monterey, California, the selected Marines learned formal Vietnamese, most often used in cities and universities. Anthony Goodrich found his formal Vietnamese training rather useless in the field. It took him six months to learn how to speak what he called "bush language," the Vietnamese slang most often used in the infantry's communications with Vietnamese.[75] The twelve-week language course also taught the history of Vietnam and its culture. As in CAP school, the Marines in Monterey learned the accepted customs for the Vietnamese. The major difference was that a grunt attending the school represented one of the very few Marines in his unit who had received this training. Beginning in 1967, the vast majority of CAP Marines received language and customs training, although it was packed into a much shorter time period.

Selected Marines in CAPs received four weeks of extended language training, offering them the opportunity to expand their Vietnamese vocabulary of military-related terms. The CAP Vietnamese language school featured two Marines trained in the Vietnamese language and five ARVN interpreters as instructors. The instructor-to-student ratio was impressive: 1:6. Over the course of the twenty-eight days of instruction, depending on the individual, Marines learned from three hundred to six hundred new words.[76] Combined action company commanders had the author-ity to choose personnel under their command to attend the school. The company commanders analyzed CAP school files, selecting those with a clean disciplinary record who had excelled in learning the Vietnamese language. Yet the efforts to bolster the language proficiency of these CAP Marines burdened the program militarily. After selection to the program, administrative processing, CAP school, and then the extended language school, some Marines had only three to four months remaining on their tour. In 1969, First CAG was unable to use the language school to its ad-vantage. With many of the First CAG Marines having slightly more than six months remaining on their tours, commanders could not afford to send them to language school for an entire month. Thus, program leaders urged their company commanders to search for Marines who also had a long time remaining on their current tour.

Another option for improving language proficiency in the villages was appointing ARVN interpreters to stand alongside the Americans in CAPs. Yet the villagers had difficulty relating to ARVN interpreters from large cities.[77] Interpreters from Hue and Saigon spoke a different dialect than the South Vietnamese villagers. Moreover, like the Americans, the interpreters were usually not familiar with the specific customs of the vil-lages. The availability of interpreters also posed a problem for CAPs. At one point in 1969, First CAG had only three interpreters for its CAPs spread across two provinces. First CAG headquarters sought to remedy the problem by increasing the number of English-speaking Vietnamese in CAPs.[78] CAP personnel in First CAG offered English classes in the vil-lages, but since their teachers had only marginal expertise in Vietnamese, quite naturally the students left with a limited understanding of English.

In CAPs, the language barrier transcended all elements of village life for Americans and Vietnamese alike. On numerous occasions, in-nocent but misunderstood dialogue between Marines and PF resulted in fistfights. Marines would unknowingly insult a PF, and the two groups would even aim their weapons at each other because of a lost translation. In a Fourth CAG platoon, when Marines called for the PF sergeant (*trung si*) Diet, his name pronounced with an American accent sounded like "anus" in Vietnamese. Villagers mocked Diet for his new nickname until the PF commander drew his firearm, threatening to shoot anybody who continued to misconstrue the name.[79] Language problems also dogged the officers in charge of the program. Before the acronym CAPs became standard, assignment in the program placed Marines in CACs, or combined action companies. The spoken sound of CAC, loosely translated, means "penis" in Vietnamese. Marines boasting to Vietnamese of their proud involvement in a CAC were commonly greeted with an outburst of laughter. Program leaders changed CAC to CAP.[80]

Another shortcoming of CAP school was that the mandatory two weeks never translated to fourteen full days of instruction. A 1969 analysis of the Combined Action Program found that slightly over 50 of the more than 122 total hours of school were spent in eating, intervals between classes, travel, orientation, graduation, and administrative duties. There were no classes on Sundays, and the last two days of the period were dedicated to testing and a graduation ceremony. When one adjusts for actual time spent in the classroom, CAP school offered only seven full working days of lecture.[81]

In a questionnaire it was revealed that the overwhelming majority of CAP school graduates (112 out of 138) deemed the language and culture aspects of training the most beneficial, which comes as no surprise considering they knew the least about those topics at matriculation. The majority of the questionnaire's respondents also believed the program should lengthen the time spent on language and culture training and drop classes on first aid and action drills. When the graduates were asked for recommendations, most respondents complained about "too much petty bullshit" like rifle and haircut inspections.[82] One CAP Marine remembers that his instructors wasted time teaching Vietnamese songs rather than what he deemed necessary: military phrases that would bolster communication with PF.[83] The program hoped that on-the-job training with the PF in the villages would build the Marines' Vietnamese vocabulary of military-related terms most often used on patrols.

The development of the selection and training processes for CAP Marines mirrored the spontaneous creation of the program itself. Reports of aggression and violence inflicted by American soldiers on civilians in I Corps necessitated both a screening system and a training regimen for CAP Marines. By the time both of these were enforced, the situation in I Corps had reached a critical juncture, with the seemingly relentless NVA assaults against Marine and ARVN units near the DMZ and the ever-concerning presence of the VC across the entire region. Considering the volatile situation, it comes as no surprise that the program was unable to maximize the potential effectiveness of these endeavors by extending the time for CAP school and eliminating the numerous loopholes that plagued the recruitment and selection process. Granted, the program's attempts to prepare these Marines for CAP duty were better than nothing at all. Two weeks of CAP school provided the Marines with some training and it also produced a quick turnover rate, placing much-needed Marines in the villages quickly. However, after only two weeks, CAP school graduates still had much to learn and many unforeseen obstacles to overcome.

Grunts coming to CAPs, their new mission to befriend the Vietnamese after their time in the infantry, found it an arduous transition. As Barry Goodson sat through the orientation session at CAP school, he was astonished. "I could not believe my ears. An American general standing there telling us that his main goal was to simply help the Vietnamese people, not to kill them! Boy, that blew the lid off everything I'd seen, heard or done before. Yes, I was a part of the action that destroyed villages, rampaged throughout rice paddies, burned huts and performed hundreds of other wantonly destructive deeds."[84] On his transition from grunt to the program, one CAP Marine marveled, "We've been up in the mountains for months, where it's been kill, kill, kill; now we come down here and are told we're supposed to love them all. It's too much to ask."[85] No two-week course could possibly prepare young Americans for life in Vietnamese villages. The Marines and corpsmen would have to make the adjustments themselves.

CHAPTER FOUR

Life in a Combined Action Platoon

In your job you will be working very close to the Vietnamese people, and you will be exposed to their culture, customs, and their ways of doing things. In this regard, my best advice is to get to know them as soon as possible, and to respect their customs and ways, as you would want them to respect yours.

—Chief of the RF-PF Division,
CORDS, III Marine Amphibious Force,
CAP graduation speech, 14 March 1970

American infantrymen roaming the countryside of South Vietnam had seen and heard the sights and sounds in the villages, but usually only in passing. Americans in the program had to realize that their assignment necessitated living with the Vietnamese and, to ensure success, adapting to their way of life. Americans arriving in a newly designated CAP village faced a population either suspicious of their intentions or petrified of the possible reprisals from the VC for interacting with U.S. troops. At least several weeks passed before the uncertainty and suspicion on both the South Vietnamese and the American side had subsided enough to foster communication, albeit limited at first. The Americans knew from their brief training that their personal lives and the life of the CAP unit hinged on their ability to collect intelligence from the villagers. Tips from villagers could forewarn the Marines about enemy whereabouts and upcoming maneuvers within their small area of operations, about two square miles. The probability of receiving accurate intelligence increased as the Americans worked toward creating amicable relationships with the villagers, including the PF. The ever-present language barrier between the Americans and Vietnamese often resulted in intelligence funneling from the villagers to PF, some of whom knew enough English to transfer coherently the information to the Marines.

As the Marines' time in a CAP village lengthened, the inevitably awkward first encounters progressed into Americans receiving invitations to villager-orchestrated dinners, weddings, funerals, and holiday celebrations. Although many Americans ultimately came to feel more comfortable in the CAP village environment, they knew that the VC was offered similar accommodations from the same civilians who acted cordially toward the Marines and corpsmen. Nevertheless, despite the Marines' knowledge that some of villagers aided their enemy, living in CAP villages brought many to a better understanding and respect for the South Vietnamese people and their culture.

This is not to give the impression that CAP villages cultivated a burgeoning and vibrant multicultural utopia. After all, the CAP Marines comprised the only foreign force in the villages during the unit's tenure of occupancy, and numerous villagers hid their support for the VC behind a friendly façade when interacting with Americans. While some Americans in CAPs established lasting friendships with the South Vietnamese, many villagers kept their distance from the Marines. Likewise, CAP Marines remained suspicious of the allegiance of particular villagers during and after their stay. Overall, however, helping civilians complete domestic tasks around the village while maintaining a secure environment assisted in creating a rapport between the Americans and South Vietnamese civilians in CAPs.

It was obvious to Americans in Southeast Asia that the Vietnamese had not progressed to the modern level the United States had achieved.[1] Until the arrival in Vietnam of the French in the late nineteenth century, the Vietnamese, particularly in the villages, had clung to the same pastoral way of life for centuries, whereas American culture thrived from advancing technological developments.[2] Many Vietnamese villages in the precolonial era experienced a high level of autonomy. The cultural, economic, and political practices—the veneration of dead ancestors and village spirits, the communal ownership of land, and the selection of village leaders—in these traditional villages remained largely untouched by government authorities. In fact, in almost all ways, the U.S. military entered a cultural environment alien to most Americans. As Frances Fitzgerald maintains, "In their sense of time and space, the Vietnamese and Americans stand in the relationship of a reversed mirror image, for the very notion of competition, invention, and change is an extremely new one for most Vietnamese."[3] Former ARVN generals Nguyen Duy Hinh and Tran Dinh Tho echo Fitzgerald's observations: "Vietnam and the United States were so dissimilar in origins, background, and civilization, and environment that they stood at the very antipodes of the human spectrum."[4]

During and after the French colonial period, traditional Vietnamese villages stood in stark contrast to the bustling, densely populated cities along the coast where the French had modernized the social and political landscape. Vietnamese villages rarely witnessed sudden sparks in population. Unlike many urban dwellers during the French and American occupations, rural villagers viewed the accumulation of wealth as antisocial, a sign of selfishness rather than success. The traditional Vietnamese family acted as the model for both village and state, forming "a crystalline world, geometrically congruent at every level."[5] To paraphrase Neil Jamieson, the traditional Vietnamese rural family comprised its own nation.[6] The family formed the centerpiece of life in the traditional Vietnamese village. The land on which the village stood was sacred to its occupants. The Vietnamese had a deep emotional attachment to ancestor worship, believing that their ancestors, buried near the village homes, roamed among them. Expressing the commitment to village and family, one Vietnamese explains, "It is . . . there, in this community of life and thought, that [one] finds the strength of the group to which he belongs."[7] Family-owned land and homes had existed for generations, inherently making them organic shrines for ancestor worship. Few villagers willingly left their homes, because to leave one's village meant parting ways with the familial and societal attachments that they so strongly embraced.[8]

When the U.S. military began arriving in large numbers in 1965, American culture gradually began to penetrate South Vietnamese cities. American cigarettes and chewing gum became staples on South Vietnamese city streets. The increasing U.S. presence also resulted in a spike of South Vietnamese urbanites attending English-language classes, giving popular American publications such as *Newsweek* and *Time* a larger reading audience.[9] The wealthier urbanites of South Vietnam and rear-echelon American military personnel assigned to cities enjoyed amenities such as electricity, television, automobiles, air-conditioners, and refrigerators. The elders of South Vietnamese cities tended to cling to the traditional Vietnamese way of life, but the younger generations adapted more to American culture. Many South Vietnamese civilians in the cities accepted employment with the American military or with U.S. private contractors and businesses.[10] Meanwhile, the more remote rural areas did not witness this infusion of American culture at the same level. Moreover, city life generally offered a safer environment during what the Vietnamese today call the American War. The majority of multibattalion-sized operations and air and artillery strikes occurred near the villages, not the cities. Only rural areas were designated "free fire zones," allowing the nearest American or South Vietnamese unit free rein to launch artillery strikes into the designated villages.

To understand why the Marines found it so difficult to interact with the civilians, one must take note of the harassment the rural villagers endured during and immediately after the French colonial period. Vietnamese peasants had suffered under foreign rule from the late nineteenth century to the mid-1950s as the French routinely violated traditional Vietnamese customs.[11] As France accelerated its modernization of Vietnam in the early twentieth century, newly built roads crisscrossed the countryside, dissecting sacred ancestral burial grounds. As these roads in turn increased automobile traffic, French motorists often struck and killed Vietnamese bicyclists transporting goods from the local market to their villages.[12] In addition to experiencing an array of cultural abuses by the French, Vietnamese peasants also fell victim to abject poverty, heavy taxes, and forced labor. To earn much-needed money for their families, many Vietnamese peasants had to leave their sacrosanct villages to work in the French-controlled industrial and commercial sectors of Vietnam.[13]

During the First Indochina War (1946–1954), French authorities assailed villages in search of enemy (Viet Minh) sympathizers. French authorities forcefully removed villagers suspected of supporting the Viet Minh from their homes to French military bases. With the belligerents vying for control of the countryside, rural Vietnamese farmers continued to tend to their lands as their ancestors had done for generations. The farmers' commitment to their lands angered both the Viet Minh, who frequented villages at night, and the French, who dominated village life during the day. The French, seeking to reduce contact between villagers and

the Viet Minh, tortured peasants who failed to comply with their inter-
rogations. The Viet Minh, frustrated by the villagers' lack of commitment
to their cause, also gained a reputation for terrorizing civilians through
interrogation and murder.[14] Much as in the American War ten years later,
villagers faced deadly consequences for openly supporting either of the
belligerents. Yet peasants who seemed indifferent faced the wrath of the
skeptical and suspicious military forces, each of which assumed they supported the opposing side.

After France's defeat in 1954 and the subsequent partition of Vietnam at the 17th parallel, two separate nations emerged, each backed by
superpowers jockeying for ideological influence in Southeast Asia as part
of the cold war. Unlike the Soviet Union– and Chinese-supported Communist state in the north, the Democratic Republic of Vietnam (DRV),
the American-backed Republic of Vietnam to the south was a fledgling
nation in the immediate aftermath of France's defeat.[15]

To oppose Ho Chi Minh and the Communists to the north, the United States supported the RVN presidency of Ngo Dinh Diem, a fervent
anti-Communist and devout Catholic who kept a firm political grip on
the government and people of South Vietnam during his reign. Plagued
with corruption and an overbearing appetite for control, Diem's government distanced itself from the South Vietnamese people, many of whom
ultimately deemed the RVN government illegitimate.[16] Diem filled RVN
government positions with his immediate family members, including
Ngo Dinh Nhu, who led the security police force. Nhu's wife, the infamous Tran Le Xuan, also known as Madame Nhu, made headlines in the
early 1960s for her insensitive references to Buddhist self-immolations in
protest of the Diem regime as "barbecues." Diem also cemented control
over the South Vietnamese military, appointing officers based on their
dedication and loyalty to the RVN president rather than their merit as
competent military leaders.[17]

Beginning in the late 1950s, Diem tried to stem the rising influence
of Communist insurgents in South Vietnamese villages. Yet, as Neil Jamieson argues, the RVN "embodied principles that related in no fundamental way to the family and village paradigms of Vietnamese culture."[18]
In 1959, the RVN relocated peasants from their ancestral homes in the
Mekong Delta to "agrovilles." These new settlements featured forced civilian labor and local security forces charged with separating civilians from

insurgents. The RVN ended the agroville experiment little more than one year after its implementation, but not before many of the dislocated civilians had already pledged their allegiance to the Communists.[19] Due in large part to the rural population's animosity to agrovilles, Diem replaced them with South Vietnam–wide "strategic hamlets," a new tactic for isolating villagers from insurgents. The start of the strategic hamlet program corresponded chronologically with the creation of the National Liberation Front (NLF), a southern-born political movement to rally the rural masses of South Vietnam against the United States and the RVN. Ultimately known to U.S. and RVN political and military officials as the Viet Cong, the NLF and its military arm, the People's Liberation Armed Forces (PLAF), had made notable gains in the villages of South Vietnam when Diem began the strategic hamlet program. Civilians constructed their own fortifications for strategic hamlets, without pay, in their villages. As with the agrovilles, Diem forced villagers to cover the costs of the strategic hamlets. Diem believed that because he was protecting the villagers, they should carry the burden of building and paying for the strategic hamlets. By April 1963, the RVN had created six thousand such strategic hamlets and planned to add several thousand more by July.[20] Ultimately, the VC infiltrated many of the strategic hamlets, with much of its success due to villagers willingly protecting the insurgents inside the fortified perimeter. For many villagers, the VC offered a more promising future, both physically and economically, than the Diem government.[21]

James Trullinger's *Village at War* examines the evolution of social and political dynamics in the village of My Thuy Phuong, southwest of Hue, from French colonial rule through the American War. The villagers of My Thuy Phuong, which ultimately housed a CAP, were hopeful that the Diem government would bring prosperity and a sense of nationalism to the village. As the Diem era in South Vietnam unraveled, the villagers lost their faith in the RVN's ability to instill stability and peace in South Vietnam. As one villager exclaimed about Diem, "Later we saw that there was no peace, and that the leaders were not honest. So how could we continue to support that man?"[22] The villagers began to equate Diem with the French, who had brought sheer misery to My Thuy Phuong.

In the two years after Diem's assassination in 1963, numerous regime changes undermined the potential for political stability in South Vietnam. In 1964, under the direction of Premier Maj. Gen. Nguyen

Khanh, the RVN again attempted to gain control of the rural population. In the northernmost RVN province of Quang Tri in I Corps, the RVN province chief Hoang Xuan Tuu began training civilians as cadre to enter South Vietnamese villages as pacification teams. The fifty-man pacification teams aimed to teach villagers self-defense during the day so they could fend off the VC at night. The teams, which usually stayed for several days, slept outside the villages at night to lay ambushes for the VC. U.S. advisors in Quang Tri believed that stationing the cadre outside the hamlet was preferable to inside village confines where the civil servants would pose as vulnerable targets for the VC. Yet many questioned the durability and determination of these civilians, who came from the provincial level with little experience in military affairs and rural customs. As one reporter opined, civil servants of South Vietnam "tended to be dead weight in the war effort."[23] The failure of the RVN to connect itself politically with the people of South Vietnam was not exclusive to I Corps. In late 1967, a JUSPAO research report found that 75 percent of the 1,313 people interviewed in IV Corps proved unable to reveal what the RVN constitution meant to them personally.[24]

CAP school had not fully prepared the Americans for the culture shock that accompanied their arrivals in the villages. "Awe and revulsion" were the words one CAP Marine used to describe his initial reaction.[25] When Sgt. Mike Murphy entered the CAP village of An Phong as its commander in 1969, he thought he had traveled back in time to the eighteenth century.[26] The program had thrust these Americans into an environment without running water, electricity, trash depositories, or toilets. Although American line units endured similar conditions while on patrol, they ultimately returned to base camps, many of which offered electricity through generators, designated trash dumps, and outhouses. As the Americans wandered around the villages, some stumbled upon civilians defecating in the open.[27] Villagers simply massed human feces into large piles near the village.[28] Other customs struck right to the heart of the young Marines' taboos. During the machismo-driven weeks of boot camp, drill instructors had attempted to humiliate underperforming recruits by labeling them homosexuals and sissies. Kyle Longley argues that the homosexual references formed a "groupthink that often marginalized individuals who appeared in any way gay."[29] When Paul Kaupus first arrived in his CAP village, seeing two Vietnamese heterosexual males holding hands, a sign of friendship, was an eye-opening experience. Coming from a culture where leisurely holding hands with another male would undermine one's heterosexual masculinity, Kaupus, although apprehensive at first, did so on occasion to avoid overtly disrespecting the villagers' customs.[30] Not all Marines accepted the practice, however. One defiantly slapped the wrist of a PF attempting to hold his hand.[31] At a CAP in Quang Tri province, Edward Palm even recalls heterosexual PF engaging in mutual masturbation.[32] Indeed, the military phrase "adapt and overcome" aptly applied to Americans in CAP villages.

The poor sanitary conditions in the villages left Marines dumbfounded. Vietnamese occupants in CAP villages bordering streams and rivers washed dishes and clothes and bathed in waterways with floating water buffalo droppings and the occasional dead body.[33] Without toilets, showers, or trash cans, the Americans in CAPs made attempts to increase the general sanitization of everyone in the village, including themselves. Soap became one of the most frequently requested civic action items from the Marines. The Americans used the soap to shower in the villages, either using the local watering hole or taking advantage of monsoon downpours. Either way, Americans taking showers always drew a large, curious Vietnamese audience.[34] Marines constantly urged the Vietnamese to clean their bodies, and on occasion took village children, mesmerized by the Americans and their hairy arms and chests, for a leisurely swim to introduce them to the foreign practice of bathing with soap. As this became routine in many CAPs, village kids bathed themselves without persuasion. Americans held health and sanitation classes on such a frequent basis that some villagers ultimately took control of coordinating their own "shower calls." For recreation purposes, the Marines also offered swimming lessons to villagers.[35]

When invited by the villagers for dinner, a common occurrence in CAPs, the Americans had to adjust to Vietnamese cuisine.[36] Although many Marines were apprehensive about certain unidentifiable food items, such dinner invitations made the Americans feel welcomed by the locals. Regarding village dinners, Igor Bobrowsky reflected, "The fact that you could be regarded pretty much the same as everybody else there was phenomenal. I was floored by it."[37] The Vietnamese also reaped benefits from entertaining Americans in their homes for dinner. The Marines donated their extra C rations to the host family in addition to providing an extra set of hands for any unfinished domestic work. The food prepared for the dinner guests aroused sanitary concerns for the Americans along with thoughts of "What the hell is that?" Yet out of respect, most Americans consumed—or at least pretended to consume—the mysterious food and drink. The Vietnamese offered whole animals as they had appeared when they were killed moments before dinner. Sgt. Mike Murphy, the commander of a CAP in southern I Corps in 1969, had apprehensions about eating a recently slaughtered hog that a family placed before him. Murphy, known as Trung-si Beer (Sgt. Beer) to the villagers for his drink preference, accepted the chunk of hog a villager aggressively ripped from the animal's carcass. Wary of the health consequences but not wanting to offend the family, Murphy pretended to chew and enjoy the food but, with the hog meat still in his mouth, covertly exited the home and hid the chewed piece of flesh in his pocket.[38] Another CAP Marine pretended to sip goat's blood from a bowl offered by one of the elder males in the village. He raised the bowl to his mouth, coating his lips with the blood to give the impression that he enjoyed the offering, but never swallowing.[39]

To comprehend how socialization between the Americans and South Vietnamese developed, one must first understand the daily routines of the CAP Marines. The Marines continually performed military and civic duties in and around the village on a daily basis. For the stationary CAPs, generally no more than three Marines stayed in the compound at all times, while the others cleaned their weapons, slept, planned for the next patrol, and performed civic action projects. CAP Marines knew that although they occupied a specific section of the village, the VC could secretly enter a temporarily unguarded location in one of the hamlets distant from the American compound.[40] A handful of Marines would accompany at least the same number of PF on daily patrols, usually planned days in advance, that lasted for several hours. CAPs established a rotation whereby every Marine participated in at least one patrol in a twenty-four-hour period.

Patrols, especially night patrols, which lasted until dawn the next day, were frightening experiences. Several Marines and most PF stayed behind in the villages for security purposes until the patrols returned. According to the popular expression from American grunts in Vietnam, "We own the day, Charlie owns the night"—the VC carried out most of its activities under the cover of night. In the hours of darkness, the social dynamics of the village changed considerably. During the day, when a relative peace characterized most aspects of village life, Marines walked through the hamlets. At night, with the perimeter gates closed and curfew enforced, the village became eerily quiet and the Marines more alert. Contrary to day patrols, at night the combined units were more cautious about the possibility of engaging the enemy in battle. Night patrols were extremely tense and emotionally draining for the Americans. A few Marines and at least an equal number of often-unreliable PF trudged through the triple canopy jungles that obscured the moonlight, not knowing what or who lay ahead of them. Reminiscing about his first CAP patrol after serving in the infantry, Ron Schaedel, then a lance corporal, admits, "I almost shit my pants. I was so scared."[41] As an infantryman, Schaedel had roamed the treacherous jungles of I Corps, day and night, with dozens of his fellow Marines. This was a frightening experience, but he had grown accustomed to it. Patrolling at night alongside only a handful of U.S. Marines was a far different prospect, for the Americans knew that at any moment an exponentially larger enemy force could overwhelm the patrol and possibly the entire CAP village. For the "fucking new guys," participating in CAP patrols was unlike anything they had experienced in their military careers. Granted, CAP duty in general differed greatly from any of their previous military assignments. However, infantrymen like Schaedel had participated in numerous patrols before landing in a CAP. They had learned how to find and kill their enemy in the bush, and they had endured the fear, frustration, and exhaustion inextricably linked with combat duty in Vietnam. Still, the nature of CAP patrols seemed to present an entirely new set of military obstacles for the newest CAP Marines to overcome. Most proved able to adapt to the new environment, but some did not.

CAP commanders had the authority to relieve any Marines in the village they deemed unfit for duty. Most CAP commanders were quick to rid their team of enlisted Marines who presented discipline problems or those who did not adhere to accepted village customs. Marines who had repeatedly undermined the rapport with the PF or civilians were usually kept on a short leash by the CAP commander. The CAP commander at Binh Nghia had to dismiss one of the original members of the unit after he had started four firefights by discharging his weapon erratically out of nervousness at the first sound he heard. Another Marine in Binh Nghia was dismissed for attacking a PF accused of stealing a radio (he claimed he was borrowing it) from a civilian's home. That Marine had shown signs of laziness on patrol and had stated his belief that his main priority in the CAP was to "shoot people." His CAP commander sent him to an infantry unit. The CAP commander also had the authority to reject incoming Marines. At Binh Nghia, the CAP commander returned new arrivals to their original outfit after one of the long-standing enlisted Marines in the village told the commander that he knew the "Fucking New Guys," all of whom had joined the program after the commander of their infantry unit kicked them out for disciplinary problems.[42]

To achieve any measurable success in CAPs, the Americans first had to address the daunting task of gaining the villagers' trust. Dinner invitations to village homes was one indication of that trust. However, one careless mistake from the Marines and corpsmen or from passing American units threatened to extinguish progress. Ron Schaedel remembers that in the village of Hoa Hiep, "it took a while" to rebuild trust with the villagers after someone from a U.S.-marked truck unaffiliated with the CAP unit threw an object from the passing vehicle, injuring a child playing nearby.[43] The late anthropologist Gerald Hickey, who worked for the RAND Corporation studying the montagnards, made several visits to numerous CAPs. At a CAP near Hue, Hickey met with the village chief, who told him that the Marines were well behaved and provided excellent security. The only complaint from the chief was that the Marines had accidentally demolished a village council house gate.[44]

It was obvious to many Americans in CAPs that the villagers did not trust them when they first arrived. Many villagers had presumed that U.S. Marines had landed in South Vietnamese hamlets only to kill civilians and rape women.[45] VC cadre repeatedly warned villagers, "The Americans will steal your possessions—kill, cripple and torture your people—take your children's food. Remember, of the Americans none are more cruel, violent and treacherous than the Marines."[46] With these fears fresh on the mind of the village chief of An Phong, he exported all women between the ages of eighteen and thirty when he learned of the impending arrival of CAP Marines.[47]

Civic action was one of the most productive means of enhancing social interaction in the villages. All U.S. military branches in Vietnam took part in distributing supplies and personnel to bolster village infrastructures and the everyday lives of the South Vietnamese. In CAPs, the Marines and corpsmen constructed schools and wells, repaired nearby roads and bridges, and delivered food and medical supplies and services to South Vietnamese villagers. The organizing and construction of civic action projects afforded the opportunity for the Americans and South Vietnamese to collaborate on projects and activities, helping to alleviate the initial social tension that permeated a village when a CAP began its operations.

From the creation of IIIMAF in 1965, the U.S. Marines employed civic action as a means to improve security for the peasant population in I Corps. IIIMAF civic action was born in part out of the South Vietnamese government's fear of losing its ability to spread sovereignty to the peasant population. With the VC constantly spoiling the RVN's pacification plans, the South Vietnamese, with the help of American military personnel, aimed to gain the allegiance of the peasant population by distributing civic action supplies. In 1965, buoyed by the early successes of combined unit experiments in I Corps based heavily on civic action, the distribution of supplies and medical attention for civilians became a top priority for IIIMAF headquarters.[48]

Even after the creation of CORDS in 1967, the process of requesting and receiving civic action materials was time consuming. At times the journey through bureaucratic channels lasted more than two weeks before the supplies arrived at their final destination. To request materials, the program designated one Marine, usually the CAP commander, as the U.S. civic action liaison for the village. After surveying the village and speaking with civilians to gauge what materials they needed, this representative sought approval for the proposed projects from the hamlet and village chiefs. Then the chiefs and the Marine sent their requests to CAG headquarters, where Marines and ARVN officers finalized arrangements for the delivery of the material. CAGs had warehouses stockpiled with the most commonly requested civic action supplies, which shortened the distribution process. If a village needed a temporarily unavailable item, personnel in CAG headquarters had to scramble outside the group compound to find the material. For example, when First CAG received a request for a wooden prosthetic leg, the civic action representative at headquarters had to travel to Saigon to find one.[49]

To avoid the bureaucratic maze, CAPs in Quang Nam initiated "No-Cost/Non-Material" civic action, a combined American-Vietnamese effort to complete village projects with neither funds nor material. This eliminated the delays in the civic action request and delivery process. The lack of material and funding necessitated smaller projects, such as digging holes for trash dumps, that required only several hours of work rather than the several days or weeks needed to construct wells and buildings.[50] The implementation of the "No-Cost/Non-Material" projects coincided with the program's transition from stationary to mobile CAPs. Mobile CAPs benefited from the new civic action approach in that the roaming units no longer had to worry about stashing material at one central location in the village. However, the shift to mobile CAPs sacrificed large civic action projects for more effective military patrols and heightened security in the hamlets. Daily patrolling to different locations to avoid predictable routines easily detected by the VC, the Marines could not spare the time needed to help with the more time-consuming civic action projects. On the move, mobile CAPs could store excess civic action supplies in each hamlet, start smaller projects, and add to those when the next patrol entered that specific location.

The Americans tended to dominate the management and construction of civic action projects. Although the "civic action handbook" distributed at CAP school warned against an American-centered civic action program in the villages, the Marines frequently finished projects without seeking Vietnamese civilian cooperation, most notably during the first three years of the program. The slow and methodical pace of the Vietnamese frustrated the impatient Americans. Vietnamese-coordinated civic action projects could last for weeks without any visible progress. Examining the cultural differences regarding the concept of time, two former ARVN generals contend that "while Americans regarded action as a compulsion and something to be performed aggressively in the shortest possible time, Vietnamese seemed to view time as an eternal commodity, an ingredient of the panacea to all problems." For the Vietnamese, slow progress on projects was not due to laziness or evasion, it "merely implied waiting for an opportune moment to act."[51] Completing projects on their own gave the Vietnamese a sense of pride. When the Americans handled the projects themselves, the villagers seemed reluctant to make use of the finished product.[52]

By 1970, graduates of CAP school had begun hearing lectures about the importance of allowing the Vietnamese more autonomy and independence with the civic action projects. One of the speakers at a CAP school graduation ceremony urged the Marines to "be patient in your dealings with the people you will work with. Don't try to ram your ideas down their throats, but rather plant the seed and they will soon do what you want them to do—thinking it was their own idea, and their decision. This is the way it works the best."[53] For the U.S. military and RVN representatives in I Corps, civic action equaled progress in the countryside. Ordered to "do some civic action," Marines concentrated on finishing projects quickly. To keep their superiors happy, even if the CAP village did not need civic action materials, the Marines kept a constant influx of requests rolling into CAG headquarters.[54]

Although assuming total control of civic action did not completely ruin any existing rapport, it certainly did not help. Many CAPs made adjustments to transfer most if not all of the civic responsibility to the shoulders of the Vietnamese. Several CAPs held classes for hamlet and village chiefs to teach them the proper procedures for attaining civic action material. The chiefs began deciphering for themselves what the village needed most and dispatching the requests to the respective CAG headquarters. Yet placing the responsibility in the hands of the Vietnamese did not always result in an improved civic action system. In the case of CAP Echo-2 in the village of Hoa Hiep, north of Da Nang, the Marine commander gave the village chief autonomy with civic action. The chief requested cement for civic action projects—and the Marines were later dumbfounded to find that he had used the material to build himself a new house.[55]

CAPs also accepted civic action support from American nonprofit organizations such as Cooperative for American Relief Everywhere (CARE) and Catholic Relief Services (CRS). In addition, the United States Agency for International Development (USAID) assisted in the civilian relief efforts of CAPs. Until the establishment of CORDS in May 1967, which placed all American pacification efforts under one civilian leader, the distribution of supplies across South Vietnam suffered from disorganization, a by-product of the decentralized nature of the civilian civic action agencies. Each agency had its own mission, logistical network, and reporting system; each pursued its own goals rather than collaborating with the others. Before the creation of CORDS, all the agencies fell under the control of the U.S. ambassador in Saigon. Yet the ambassador had scant control over the agencies. According to William Corson, CAPs experienced trouble in procuring useful materials from USAID. The program created alliances with several USAID representatives who wanted to help CAPs by "counter-looting" in response to corrupt RVN officials who pocketed USAID funds and material rather than distributing them to the villages.[56] Ron Schaedel's CAP received corn flakes from USAID. "We didn't have any milk," Schaedel recalls. "Why the hell do we need corn flakes?" Schaedel and his fellow CAP Marines gave the corn flakes to the village pigs.[57]

Throughout the war, IIIMAF attempted to bring organization to the civic action program in I Corps. In 1965, to boost efficiency, IIIMAF commander Maj. Gen. Lewis Walt created the Joint Coordinating Council. Consisting of representatives from all the civic action organizations in I Corps, the JCC synchronized the management and distribution process under one administrative umbrella. Representatives from every institution involved with civic action formed numerous committees, each dedicated to specific functions such as public health, education, roads, and commodities distribution. In late 1966, the structure of the newly created Office of Civil Operations (OCO) in Saigon resembled that of the JCC. The OCO brought the numerous pacification programs in South Vietnam under the leadership of ARVN major general Nguyen Doc Thang, who controlled the Ministry of Revolutionary Development and in his new role reported directly to the U.S. ambassador in Saigon. The OCO placed regional directors in charge of each of the four corps tactical zones in South Vietnam, but the new program still suffered from the various agencies acting independently of one another. To finally halt the disorganization, the United States created CORDS in 1967, which placed one American civilian in charge of pacification. The former OCO corps directors inherited the same position in CORDS. In I Corps, the new CORDS director had an easy transition to his assignment since the JCC had for two years laid the foundation for a coherent, organized pacification system.[58]

All four CAGs supplied their CAPs with recreational equipment, which provided a temporary cure for the Americans' boredom while also presenting more opportunities for interaction with the villagers. CAG headquarters distributed horseshoes, darts, playing cards, Monopoly, and checker sets (to name a few) to their respective CAPs. Numerous CAPs used gaming supplies to start village chess clubs and play cards with the villagers, games that some Marines admittedly let the South Vietnamese win.[59] Second CAG made musical instruments available at its headquarters in Quang Nam and hosted talent shows that showcased the musical abilities of Americans and Vietnamese from the group compound. In 1970, CAPs in Quang Nam received visits from the Hoi An–based cultural drama team, a group that entertained villages with songs, skits, and magic acts laced with pro–South Vietnam propaganda. During the festivities, many of the village elders invited Marines to their homes for dinner that night. With a South Vietnamese revolutionary development team leading the festivities, the CAP Marines at Binh Nghia participated in a night of entertainment for the villagers. Nearly all of the five thousand villagers enjoyed musical presentations from a quartet of CAP Marines who gave their best efforts to perform American and Vietnamese songs, much to the amusement of the crowd, which called them back for numerous encore performances—despite their obvious lack of talent.[60]

American-centered entertainment in the villages often included Hollywood movies, particularly westerns and comedies, courtesy of a borrowed projector from the nearest army or Marine line unit. The villagers demanded a rewind several times to gawk at the panoramic shots of large American cities. The female villagers watched in awe when scenes of Hollywood-created American kitchens graced the screen. Although many of the villages lacked access to electricity, some CAPs bartered with American units outside their areas of operation for generators. One CAP traded captured enemy weapons and supplies to nearby Navy Seabees in return for a hundred-volt generator. After obtaining that, the Americans soon acquired lamps, a record player, and a television from the nearest PX. However, the generator created numerous problems, starting with the noise. Moreover, Marines began scheduling their patrols around television shows. Four weeks after trading for the generator, realizing that it was a detriment to the overall military effectiveness of the CAP, the Marines traded it for timber to build a dispensary and a small village office.[61]

Americans took part in village holiday celebrations. Although the majority of Vietnamese practiced Buddhism, villagers and PF helped the Marines decorate the main roads and trails near the villages with Christmas decorations provided by American organizations or mailed from the Marines' families. In return, the Marines helped the Vietnamese adorn the same trails and roads for the Tet holiday. The minority of Catholic

Vietnamese families in the villages invited Marines of the same Christian faith to their homes for Christmas lunches and dinners.[62]

Some of the Americans established relations, in some cases sexual, with female villagers. Village elders frowned upon women who voluntarily engaged in sexual relations with the Americans. A Marine had to get approval from a woman's parents and brothers before he could sleep with her. To accomplish this required "language skill, facial charm, diplomacy, long patience, and much luck." West reports that three Marines in his CAP had long-standing sexual relations with village females, and in doing so risked losing the villagers' respect if the woman's parents complained to the hamlet or village chief.[63] If the Americans attempted to import a Vietnamese prostitute to the village, civilian elders vehemently protested, in some cases physically throwing her out.[64] Edward Palm contends that in his CAP, the Marines were insensitive to the standards of sexual behavior in the village. Villagers were visibly upset when the Americans attempted to speak with unmarried females before the proper customary introductions.[65]

The corpsmen were the most indispensable members of CAPs. The daily medical services the corpsman, or "Doc," offered to villagers provided the vital social link between the Americans and South Vietnamese. If his reputation among the villagers as an effective Western doctor took a hit, the entire CAP suffered.[66] The corpsman conducted MEDCAPs daily at a central location in the village. MEDCAPs provided medical and dental treatments, and in general promoted sanitary health conditions. Villagers lined up by the hundreds to receive treatment for everything from headaches to life-threatening injuries. Chuck Ratliff shares the thoughts of his fellow CAP Marine veterans: "Doc won more hearts and minds than all of us combined."[67] In Bruce Allnutt's report on CAPs, most of the nearly three hundred villagers interviewed credited MEDCAPs as one of the most beneficial components of the entire program.[68]

The origins of MEDCAPs in Vietnam reach back to the U.S. advisory period.[69] Between 1963 and 1971, U.S. military medical personnel conducted 40 million MEDCAPs throughout South Vietnam. The United States first implemented MEDCAPs as part of the overall push to use civic action as a tool to penetrate the countryside and connect the peasants with the South Vietnamese government.[70] Beginning in 1965, in hopes of making that political connection a reality, South Vietnamese medical personnel accepted responsibility for the MEDCAP program, with Americans standing by as advisors. Stationed at a naval hospital at Hoi An south of Da Nang, Stanley Bloustine highlights the problems he believes accompanied Americans during MEDCAPs. American medical personnel dispersed pills and candy to children but never considered integrating the Vietnamese health care system. "I'd try to convince them our job was to convince the people that the Vietnamese government could take care of them," Bloustine recalls, "not that we could take care of them."[71]

Americans arriving during the latter advisory years took note of the unfathomable medical and sanitary problems that plagued Vietnam, especially in the rural areas. Typhoid fever and cholera ran rampant in villages. In 1963, large garbage dumps near populated areas attracted fly and rodent populations, providing fertile breeding ground for the plague. In one CAP alone during December 1968, the corpsman treated more than two hundred persons infected with the plague.[72] Polio had also found its way to South Vietnam, and in the rural areas without electricity, the lack of refrigeration quickly rendered the oral vaccine useless.

The general absence of trained medical personnel in I Corps and South Vietnam made corpsmen vital to the program's viability and success. In 1969, only thirty-six doctors and three dentists existed for the 2.5 million inhabitants of I Corps.[73] During the summer of 1965, Congress investigated the RVN-managed medical aid for civilians. Of the twenty-eight provincial hospitals in South Vietnam, only eleven had working surgical units because of a shortage in medical personnel. A congressional subcommittee had found that by March 1967, South Vietnam was suffering an average of one hundred thousand civilian casualties per year. Part of the blame for the high rate rested in the fact that most of the forty-three provincial hospitals in South Vietnam by October 1967 did not have electricity, drinking water, or sanitation facilities. At some locations, civilians in need of surgery had an estimated wait time of one year because of the shortage of physicians. Although the United States poured $37 million into the civilian medical budget in 1967, facilities for civilians "were totally inadequate to meet minimum needs of the country in time of peace, much less in time of war." The medical situation in 1968 continued to disappoint Congress, with reports of human excrement on the walls of some hospitals, and in a Quang Ngai (southern I Corps) facility, penicillin shots had found their way to the black market.[74] The abhorrent state of South Vietnam's medical facilities coupled with the scarcity of doctors made the mobile American medics and corpsmen invaluable assets for the U.S. war effort, especially in the rural areas surrounded by adverse terrain and narrow dirt roads.

After joining the program, the medical corpsmen attended CAP school with the Marines. Although they attended the same classes and lectures as the Marines, corpsmen sometimes took a short leave of absence to observe routine MEDCAPs near Da Nang. Wayne Christiansen recalls one of these led by a Vietnamese nurse: the woman prescribed pills for headaches and charcoal and mint for stomach problems. "I had no idea what she was doing, and neither did she. It was a good example of what not to do in our villages."[75]

The corpsmen had medical duties to perform each morning in the CAP villages. The U.S. Navy supplied every combat corpsman with Terramycin, malaria pills, aspirin, and bandages. Every corpsman had a logistical network separate from the Marines' civic action supply chain from which he procured medical supplies. They submitted monthly requests to their respective CAG headquarters, which on average distributed those supplies every two weeks. CAGs based the amount of resources and funding allocated for MEDCAPs on the number of treatments a corpsman reported over the previous two weeks.

The U.S. military continued to operate MEDCAPs during the war, but the non-CAP variety differed in many ways from what corpsmen in CAPs were doing. By 1966, each battalion of the U.S. Army's 173rd Brigade had at least one MEDCAP program. Even during combat operations, the unit executed MEDCAPs within its area of operations with small four-man teams: a physician and three medics. Arriving shortly after the 173rd Brigade, the army's First Cavalry Division also adopted a MEDCAP program. One brigade from the division operated a medical dispensary where local civilians could receive supplies and medical attention. However, these MEDCAPs were not stationary. Moreover, the units did not provide daily MEDCAP services like the CAPs. As medical patrols increased with U.S. military escalation, the U.S. Army noticed the problems with roaming MEDCAP patrols, which in some cases resulted in seriously wounded civilians getting brief treatment after their wounds had gone untreated for several weeks. Noticing this problem in 1967, the army's 199th Infantry Brigade established permanent treatment sites in its area of operations, open two days a week for civilians. Although this increased the number of treatments given to Vietnamese civilians, patients still had to travel as far as ten kilometers for medical aid. An army report in 1968 reexamined its medical procedures and called for monthly follow-up MEDCAP visits to a given area.[76] The CAP corpsmen proved more consistent in their treatments with daily MEDCAPs. Moreover, since the corpsmen lived in the villages, they could immediately treat any unexpected injuries sustained before or after the daily MEDCAP.

During the war, corpsmen in training at Camp Pendleton's Field Medical Service School learned about the military and political aspects of MEDCAPs. The Field Medical Service School taught corpsmen that at the battalion level, the main intention of MEDCAPs was not to improve the general health of the population. If the corpsman ever doubted whether a particular medical case had a cure, the Field Medical Service School discouraged him from evacuating that person for outside medical attention. If the corpsman evacuated the infected person to the rear, and the civilian returned to the village without a cure, the navy feared that "unwanted political damage" would ensue. According to instructors at the Field Medical Service School, the villagers would then prove reluctant to accept any further medical attention, which would diminish any hopes of gaining that person's and ultimately the entire population's allegiance. Only medical cases that would have an immediate political impact on the village (harelip and cataracts, for example) were acceptable for evacuation.[77]

When not performing MEDCAPs, CAP corpsmen frequently handed out aspirin for headaches, delivered babies, and tended to wounds sustained from stray friendly and enemy fire. In one extreme case, CAP corpsman Gary Evins performed a tracheotomy with a ballpoint pen.[78] Without knowing, corpsmen sometimes cured ailments for VC who had disguised themselves among the hundreds of noncombatants waiting in line for treatment. After stitching up a wound for a young village boy, Wayne Christiansen found out later that U.S. forces arrested the kid for making booby traps for the VC.[79] When treating the civilians who knew little or no English, the corpsmen could easily identify recurring ailments like stomachache (*dau boom*) and headache (*dau dau*). Armed with limited medical supplies and only a basic knowledge of Vietnamese medical terms, the serious, life-threatening injuries posed the largest hindrance to the corpsman's effectiveness.

The importance of the corpsmen rests in the fact that they possessed an irreplaceable skill set in CAP villages. If a corpsman fell ill with malaria or was wounded or killed in battle, the Marines and the corpsman's village counterpart (*bac si*) inherited all medical responsibilities. Many CAPs would operate for days without a corpsman. During the reporting period for August 1969, seven out of the thirty-one CAPs in Third CAG reported not having a corpsman. During one month in the summer of 1969, six of the eighteen CAPs in Fourth CAG operated for brief periods without a corpsman.[80] The absence of a CAP corpsman made an already difficult task seem insurmountable. To offset the difficulties in a CAP without the corpsman, "Doc" taught some of the Marines the basics of treating sick and wounded civilians. Yet the Marines and the bac si had neither the aptitude nor the knowledge to perform medical functions effectively. Aside from the occasional crash course from the corpsmen on how to treat sick and wounded civilians, the Marines had no prior medical training whatsoever. As for the bac si, he had very little, if any, training before the arrival of the corpsman to the village. The vast majority of a bac si's medical training occurred on the job with the corpsman. CAPs often reaped enormous life-saving benefits from the South Vietnamese medical trainees, while others had negative experiences. One CAP Marine discovered that the bac si in his village was selling drugs on the black market.[81] Kept busy accompanying Marines on patrol and treating the sick and wounded in the villages, corpsmen had little time for one-on-one training with the bac si. Moreover, being a bac si required long hours, hard work, and no pay, which accounts for the results from a program study showing that most Americans in CAPs believed the trainees had lost interest in their jobs over time.[82]

The deeply embedded cultural dynamics of Vietnamese medical practices posed perhaps the most perplexing problems for the corpsmen. Traditional Vietnamese culture centered on the family and home, making leaving one's village, even in life-threatening situations, extremely difficult. As Gerald Hickey explains, "To the villager *nha* [house] is not simply a physical structure; it is a symbol of family and hearth."[83] Vietnamese with serious injuries were often reluctant to leave their homes for outside, Westernized hospitals. If a villager required evacuation, family members often accompanied him or her to the outside hospital. But many Vietnamese villagers mistrusted Western hospitals and medicine. Traditional, rural Vietnamese families had relied on their own personal remedies for sickness. In the more remote areas of the Central Highlands, civilians for centuries had trusted the spirits invoked by local sorcerers and shamans to cure ailments. In a CAP near Chu Lai, village elders cut short Jack Broz's attempts to cure a PF's fever, telling the corpsman that his medicine was for Americans only. The elders proceeded to employ their own remedy: heating shards of broken glass and cutting small incisions down the spine of the patient, then applying an unknown liquid to each cut. They then dropped a bird's nest, a spider web, and a wasp nest into boiling water and made the PF drink the concoction. For the final step, the elders placed their patient on a bed over a bucket of hot coals. As Broz recalls, the sick PF "was no better by morning, so I called for a medevac for treatment in the rear. About a week later he rejoined our CAP."[84] CAP corpsman Curtis Englehorn wrote to his family that the local Vietnamese doctors "prescribe chicken shit as a medicine for cuts and bruises. After a week or so they see it won't work so they come to us. By that time infection is well set in."[85] The Vietnamese villagers rarely tended to serious wounds, and a villager's death in an outside hospital justified their distrust of Western medicine.[86] Shortly after its creation in July 1968, Fourth CAG headquarters noted the immense frustration in trying to convince the Vietnamese to discard their primitive methods of medical treatment. Fourth CAG connected the villagers' reluctance to accept modern medical treatment with a general "lack of assets, education and the traditional Vietnamese ability to be happy and have a good time regardless of the environment."[87] Fourth CAG's simplified explanation clearly falls short of capturing the complexity of traditional Vietnamese culture. This highlights how important it was to send CAP Marines and corpsmen to live in the villages. The enlisted troops and NCOs in the villages were able to analyze, learn, and adapt to the cultural environment on a daily basis. CAG officers, who rarely visited the CAPs, operated in relatively safe compounds far from the villages. Thus, they were less likely to comprehend the cultural complexities of the traditional Vietnamese village.

Although assigned for medical purposes, the corpsman frequently augmented the American military component in CAPs. Classified under the Geneva Convention as noncombatants, corpsmen could not actively participate in skirmishes and firefights. Yet with only a squad of Marines, CAPs often needed military assistance from all capable personnel; CAG headquarters recognized that having corpsmen participate in combat was necessary at times. Corpsmen were equipped with a pistol and sometimes M-16 rifles and M-79 grenade launchers. They patrolled alongside Marines and PF, participating, if needed, in any ensuing firefights. Lacking the experience and military prowess and knowledge of the Marines, many corpsmen received on-the-job training like the PF. Out on his first patrol in a CAP near An Hoa, corpsman John Daube found himself engaged in combat. He remembers one of the Marines shouting for Doc to fire at the enemy's tracers. Recalling that first taste of combat, Daube recounts, "When it was finished my pants were wet and soiled. But it felt good to flip an M-16 on full automatic and let it rip. It was a feeling of power and fear at the same time."[88] After a few years of the program's existence, it was commonplace for Doc to be the most experienced American in combat if transferred from an existing CAP to a freshly created unit with "green" Marines.

After the Tet Offensive, the program began to shift CAPs from a static to a mobile role. Prior to Tet, all CAPs operated from static positions within the villages. Each static CAP featured a compound at a central location in the village with living quarters, bunkers, and a sick bay. In 1968, as part of an effort to maximize security for the villages with the upcoming withdrawal of American forces, the program ordered CAPs to change from static to mobile positions within their area of operations. Mobile CAPs did not make as much use of the compound. Rather, they roamed among all hamlets in each village, using temporary bases at locations within their area of operations, including houses when this was approved by the hamlet or village chief. The CAGs encouraged the Americans to choose the least aesthetically pleasing homes. According to Second CAG sources, the enemy knew that Americans appreciated comfort and luxury.[89] The mobile CAPs constantly changed the patrol routes in and around the village, creating an erratic pattern to confuse the VC. The idea behind mobile units was to give the impression that "the CAP seems to be everywhere, but never predictably anywhere."[90] Static CAP units frequently established predictable patrol patterns that in retrospect undermined security in the village. The Marine and PF mobile patrols relocated positions multiple times per day, gaining a better knowledge of the terrain within their area of operations. More than twenty days passed before the Marines and PF returned to their initial embarkation point. Mobile CAPs did have to sacrifice lengthy civic action projects for security. Constantly on the move, mobile patrols spent less time on the larger, time-consuming civic action projects such as constructing wells and buildings, which had become popular in static CAPs. However, mobile CAPs did allow for interaction with a larger percentage of the village population and for the completion of smaller civic action projects. Militarily, as reported by the program, security within the villages improved when CAPs switched from static to mobile. While mobile CAPs did perhaps succeed in improving security, the near annihilation of the VC during the Tet Offensive definitely played a role as well.

The shift to mobile CAPs coincided with a drop in the level of enemy activity in the villages. The peak of operations conducted by CAPs came in May 1969, thereafter steadily declining into August.[91] Moreover, the increasing percentage of pacified territory in I Corps as reported by FM-FPAC, and the influx of army mobile advisory teams and Marine mobile training teams, lessened the training burden for CAPs. Mobile CAPs also resulted in a downward slide in the number of MEDCAPs offered. Corpsmen in mobile CAPs carried a limited number of supplies in their backpacks, and on the move, they simply did not have the extra hours every day to render medical aid at a central location in the village.[92]

Numerous CAG command chronologies note that CAP villagers remained loyal to the South Vietnamese government during the war.[93] Lewis Walt also argued that U.S. Marine pacification helped to create a bond between the Vietnamese people and their government.[94] However, testimony from Americans who served in CAP villages presents opposing arguments. The Marines discovered that most villagers cared little about the Saigon government. Nor did they cling to the political ideologies of either of the belligerents. Shawn McHale's fascinating work on the impact of the printed word on Vietnamese culture shows that in the decades before 1945, the success of Communism in Vietnam was not because of any appeal that typical Marxist rhetoric such as "class struggle" had on the population. Rather, Communism gained popularity when its leaders began to substitute what the Vietnamese viewed as "strange and obscure" Marxist rhetoric with dialogue that suggested a broad nationalist opposition to French rule.[95] Two decades later, with Ngo Dinh Diem leading the RVN, the rural Vietnamese cemented their loyalty to the VC when the South Vietnamese government had enacted policies detrimental to village life. Douglas Pike argued that when the South Vietnamese villager viewed the RVN with contempt, the VC became "a real and valuable thing protecting him and his village."[96] Even some members of the VC chose to fight against the RVN and the United States for nonpolitical reasons. Truong Nhu Tang's memoir of his service with the VC shows that the insurgent group encompassed persons with nationalist rather than Communist ambitions, who sought to rid South Vietnam of foreign influence and control.[97] Loyalties rested with families, not with the government of South Vietnam or a particular leader in Saigon. Villagers who had not joined the VC tended to support whichever belligerent gave their family and village the greatest chance of survival, a pragmatism many CAP Marines understood and accepted.[98]

However, as the American military presence in and around the villages increased, those villagers both committed and uncommitted to the VC shifted allegiance to the belligerent that appeared stronger at any particular time. For example, in the village of My Thuy Phuong in I Corps, when the U.S. Army's 101st Airborne Division constructed Camp Eagle next to the village in 1968, the number of villagers committed to the VC decreased significantly. This did not translate into a dramatic decrease in VC proselytizing, nor did it mean that more villagers began supporting the RVN. The arrival of the army base, which brought more disruptive and destructive activity to My Thuy Phuong, did force villagers to stray from interaction, and in some cases commitment, with the VC for fear of reprisals from the American GIs at Camp Eagle.[99]

Retrieving intelligence from villagers who had endured decades of broken promises and constant disruptions to their everyday lives would prove a daunting task for the CAP Marines. Corson estimated that six months passed in the average CAP before villagers approached the Americans with intelligence.[100] Village elders possessed numerous loyalties, some harkening back to the war with the French, when many civilians had supported the Viet Minh. Some twenty years later, numerous villagers still kept those loyalties to the new "liberators," the VC. Although a seemingly Herculean task, gaining effective intelligence was critical in creating and maintaining security in the villages. A Marine grunt who ultimately joined the program exclaimed that the infantry mostly wandered around "like an elephant" without any purpose, failing to gain any useful intelligence from the local Vietnamese. In CAPs, however, Marines viewed the Vietnamese as vital links not only in gathering intelligence but also in staying alive. Failure to retrieve intelligence increased the probability of death for the Americans. As Igor Bobrowsky put it, "If you didn't have the people on your side, you might as well blow your brains out."[101]

Marines rarely received intelligence directly from the villagers. Fear of reprisals from the VC for divulging intelligence to the Americans forced willing villagers to deliver information in a covert manner. Villagers could detect when the enemy was about to make an attack on the CAP. Vietnamese civilians also knew which villagers, in some cases within their own families, supported the VC. Marines often had to rely on detecting subtle shifts in mood among villagers or interpret certain movements that could possibly foreshadow an upcoming attack. The VC had gained a reputation for threatening to kill villagers suspected of collaborating with the Americans or of not complying with their demands. A captured VC document during the Vietnamization period of the war urged the people of South Vietnam to dismiss CAPs as "countering the interests of the US People," claiming that the Marines and PF "trample on the sovereignty of the Vietnamese nation." The document continues, "Refrain from patrolling and setting ambushes to massacre the Vietnamese people. Remain in place and let the people go freely to farm and earn their livelihood. If you keep such an attitude you will be welcomed and safe until you return home with your beloved ones. Otherwise, if you keep on doing harm to the Vietnamese people, you will be severely punished by the People's Liberation Armed forces and even by the puppet troops who are working at your side, since they are also Vietnamese."[102]

The VC often carried out such threats. In a CAP near Da Nang, the VC killed a boy after he had stolen their weapons and given them to the Americans.[103] A female teenager near the CAP at Binh Nghia befriended the Marines in a nearby line unit, making extra money washing their clothes. When they learned about her actions, a VC assassination squad entered her home at night and shot her in the head in front of her parents. The same squad entered the home of one of the hamlet chiefs in the Binh Nghia CAP, repeatedly stabbing him in order to get information about Marine patrols in the area.[104]

Although the VC frequently fed village civilians inaccurate information, the Americans often received accurate intelligence from the villagers, helping to root out VC cadre within the villages.[105] A villager in CAP Echo-

2 told the Marines the precise time the VC planned to destroy a nearby bridge. That bridge indeed imploded exactly as the villager had revealed.[106] Of the twelve operating district intelligence centers in I Corps in 1969, eight reported "constant and rapid information" from CAPs that proved "highly effective" and "of extreme value."[107] Villagers did not always cower from openly identifying enemy combatants. In 1967, after three months of living in self-isolation from the Americans, several female villagers repeatedly screamed, "Viet Cong!" to warn of a nearby enemy combatant.[108]

The children, more than any other demographic in the villages, accepted the Americans, and the Marines returned the favor. Corson believed, "The genuine feelings the Marines display towards the hamlet children is probably the most significant factor in bridging the gap between East and West."[109] Although perhaps "bridging the gap between East and West" is an overly optimistic characterization, the Americans and village children in CAPs did create emotional bonds. In the mornings when the adults began preparing for their daily duties in the fields, they rarely stopped to acknowledge the Marines. Yet the kids swarmed the Americans with friendly greetings, pinching the hair on their arms and eating breakfast with them, all while practicing their English. CAP veterans today testify to how easily the village children learned English. Not only did this assist in creating bonds of affection between the Americans and children, the English-speaking kids became a useful source of intelligence when their parents proved unwilling to divulge information.[110] As the village children congregated around the Marines, the kids innocently disclosed information about villagers they had overheard from their elders' conversations. Moreover, as the older Vietnamese witnessed the joy the Americans brought to their kids' faces, parents initially reluctant to speak with the Marines began to communicate with them. "Get a smile from the kids," one CAP veteran says, "you get a smile from the parents."[111] As many CAP veterans recall, winning the "hearts and minds" of the kids led to better rapport with the parents.[112]

"Adopting" Vietnamese kids was a common occurrence in CAPs.[113] Kids would sleep next to the Marines on their cots, eat C rations with them in the morning, and walk with the "parent" Marine to school. Quang Nam CAPs launched a "little brother" program, whereby the Marines adopted young boys from their villages and provided any domestic assistance he and his family needed. The "little brother" program led to the creation of an intervillage youth baseball league among CAPs. On several occasions, the little brothers from one platoon in Quang Nam played against the kids from a nearby CAP. Another CAP from Second CAG created a military marching squad with their little brothers. The kids proudly showed off their newly learned military maneuvers in the village. In Quang Tri, CAP Marines organized a Buddhist Boy Scouts group. Participating village youngsters held meetings complete with religious classes, games, and contests. As a testament to the emotional attachment created between the kids and Americans, village children often received extra clothing from Marines who had written home to their families to request it.[114]

On 25 August 1967, a picture of an American and a boy in the CAP village of Hoa Hiep graced the cover of *Life* magazine. The featured article, "To Keep a Village Free," documented the daily activities of the Americans and villagers in Hoa Hiep. As a symbol of the program's purpose, the cover picture shows an American with his rifle in one hand and fishing poles in the other, walking alongside a village boy on crutches. Ngo Cuoc, the thirteen-year-old boy on the cover, whom the Americans in the CAP called "Louie," lived with "hopelessly deformed and atrophied" legs. Before the CAP Marines' arrival in Hoa Hiep, U.S. military personnel had arranged for Louie to visit doctors in the United States, but the young boy returned to his village after U.S. medical specialists concluded that his crippled legs were beyond repair. However, Louie's trip to America did improve his English-speaking skills: he addressed everyone in the CAP as "sweetheart." Louie ran errands for the Marines and served as an interpreter, making him one of the most beloved and protected of all the villagers. According to the journalist who covered the CAP, "Among the other, healthy youngsters in the village, Louie is the man."[115]

Although the Americans generally had favorable experiences with village children, some of the younger civilians used their friendliness to achieve goals for the VC. Near An Hoa in 1969, as a Vietnamese boy approached a CAP village with a backpack, the Marines noticed a wire strung from his back down to his arm. One Marine shot the boy. As the child fell to the ground, he pulled the wire, igniting the explosive device in his backpack. The boy's parents were VC who had sent their son on a suicide mission to kill as many Americans as possible.[116] Bill Bennington recalls a similar incident involving a Vietnamese girl who for months would ride her motorbike to the outskirts of the hamlets to sell ice cream and soda. Just as the kid had gained the Americans' trust, she rode into the village one day and detonated an explosive device hidden under her clothing.[117]

Although Marines may have grown more comfortable with their social and cultural surroundings, they were always concerned that some of the civilians supported the VC. Even after returning home to the United States, many CAP Marines continued to wonder if particular villagers secretly assisted the VC. Marines in the village of Binh Nghia suspected but never confirmed the allegiance of villagers who initiated peculiar sounds and movements during CAP patrols. During the evening hours, the Marines believed, some of the villagers would warn the VC in the outlying areas of their upcoming patrol. As the patrol mobilized, one civilian male coughed "loudly and falsely," while another woman transferred the lantern in her home from one location to another whenever the Marines passed near her home.[118] Often the village women would corral the kids shortly before the VC initiated a firefight, a sure indication that the civilians knew the enemy was approaching.[119] Yet the Marines could not risk the consequences that interrogating the suspected villagers could have on the subsequent patrol or the rapport they had established with the people. As happened so often in infantry units, South Vietnamese, whether they supported the VC or not, would surely support America's enemy if the United States acted too aggressively. CAP Marines ordered villagers suspected of being VC supporters to remain in their homes after curfew and declared they would shoot anyone attempting to leave after dark. In a CAP in Quang Tri, the Marines demanded that the villagers keep their dogs inside at night to safeguard the outbound patrol from barks that could alert VC in the area to their movements.[120]

Some actions taken by the Americans prove their uncertainty about civilian loyalties and intentions. The Americans rarely walked around the village, day or night, alone; they went in pairs. When receiving a haircut from local Vietnamese, Marines always kept a .45 pistol on their laps in case the barber tried to use his or her razor as a weapon.[121] When former CAP Marine Tom Harvey ventured back to Vietnam in 1989, he reconnected with a woman whom he had befriended during the war. In their correspondence after that visit, she revealed that she and her uncle had supported the VC during the CAP's tenure. She also admitted that her uncle participated in firefights against Harvey and his fellow CAP Marines.[122]

Ron Parks remembers finding out after he departed his initial CAP that one of his best village friends had been a member of the VC. After the program transferred Parks to another CAP, he wrote a letter to his previous village in hopes of reaching his Vietnamese friend. The letter came back informing Parks that his friend had been captured as "confirmed VC." Reflecting on this particular event, Parks writes that some of the villagers, including his friend, "had to lie to us to protect their families."[123]

Generally, CAP Marines were not naïve about the reasons behind villagers' ever-changing allegiances. Some even sympathized with the precarious and dangerous situation the war presented the villagers. "During the day they were on our side, and during the night they did what they had to survive," former CAP commander Mike Murphy says. "And I don't believe a CAP Marine alive would hold that against them."[124] Interviewed while stationed in his assigned village, one Marine echoed Murphy's sentiments, opining that the Vietnamese villagers "could be real prosperous people if they had half the chance. For hundreds of years some foreign country has been in here bleeding their resources."[125]

James Trullinger estimates that before the American military's arrival in Vietnam, at least 75 percent of the villagers of My Thuy Phuong supported the NLF, with only 5 percent committed to the RVN and the remaining 20 percent uncommitted. In 1965, when the Marines created one of their enclaves around Hue, just northeast of the village, the pro-RVN group in control of the land surrounding My Thuy Phuong was the only element preventing the village from having the status of a "V.C. village." From 1964 to 1967, the NLF capitalized on the RVN's weak political influence in the village. During those three years, NLF cadre rhetoric focused on opposing the Americans, promising a better life through sacrifice and struggle within the village. A peasant from the village commented that the NLF "taught us that everyone had to cooperate, and that we were all struggling for freedom and independence."[126]

Overcoming the language barrier was one of the most difficult obstacles for the Americans. Most could articulate simple questions and answers in Vietnamese, but attempting to explain something complex presented problems.[127] Any complicated communication often had to filter through civilians, kids, assigned interpreters, or the PF. Some villagers, knowing the Americans' rudimentary knowledge of the language, slowed the delivery of their speech to allow the Marines to piece together enough familiar words to give them a general idea of the message's import. As the Americans' time in the villages lengthened, their language skills gradually improved. Learning the language was an even more formidable obstacle for those who entered villages before the formal introduction of CAP school in 1967. Yet Ron Schaedel relates that despite never having received language training, he and his fellow Marines gradually increased their Vietnamese vocabulary through speaking with villagers. The Americans began to learn key words and phrases and by the end of their tours could piece together more complex phrases and sentences.[128]

Ironically, VC who had defected to the South Vietnamese and American side became invaluable interpreters for CAPs. In 1963, Diem enacted the *Chieu Hoi* (Open Arms) campaign in South Vietnam, aiming to rally VC defectors to the side of the RVN/United States. From its inception until January 1971, the Chieu Hoi campaign corralled more than 190,000 defectors.[129] Derived from the insurgency in the Philippines during the 1950s, defectors, known as Hoi Chanh in Vietnam, were offered amnesty rather than imprisonment or death. Every province in Vietnam contained a Hoi Chanh reception center, where South Vietnamese police interrogated them to extract any useful intelligence and possibly identify ARVN deserters or bogus defectors acting as double agents for the VC. Upon completing the interrogation process, the Hoi Chanh received an RVN identification card, a food allowance for himself and his family dependents, clothing, and a monthly salary for his upcoming services. The defectors also received monetary rewards for enemy weapons they found. After a three-day RVN indoctrination course, the Hoi Chanh could choose to establish residence in an urban area, a designated Chieu Hoi village, or his home village. Although RVN placed many of the Hoi Chanh in South Vietnamese military positions, including the territorial forces, some found employment in the Kit Carson Scout Program; these were assigned to a Free World Military Force (FWMF) unit to provide expertise on the geography and tactical tendencies of the VC.[130]

Numerous CAPs received Kit Carson scouts indigenous to the village in which they operated. The Hoi Chanh proved to be a tremendous military resource for CAP Marines, providing unprecedented access into the tactics and likely location of the VC. The scouts could easily identify the VC's favorite hiding spots and the most often used shortcuts around the village. They could also help identify previously undetected local VC cadre.[131] Kit Carson scouts were also a great boon to communication in the CAP villages, acting as interpreters between Marines and villagers.

As the Americans' time in their CAP villages neared an end, many had gained a deeper respect for the Vietnamese people. In early 1967 Sgt. James White left Binh Nghia as the CAP commander. The Marine's family received letters from the villagers before he had even arrived home. The village chief wrote to White's parents, "My people are very poor and when to see a marine they are very happy. When V.C. come to people, people come and talk to Sgt. White so Sgt. White can talk to P.F. and marine to fight V.C. Maybe die." The White family also received a letter from a Binh Nghia schoolteacher who said that White

> had done a number one job. For our people I want to thank you for having a number one son. About 3 months ago my village was having trouble with Viet Cong and Sgt. J. D. White and Sq. help protect my people and land. I want to thank him very much for helping have peace in my village. I'm very happy that Sgt. White is going to home. I wish in my heart that every man was like him. I hope in my heart that Sgt. White does come back when my country is at peace. Many of my American friends have died. I'm very sorry at has happen to your people. I hope some day we will all have peace and Charity.[132]

Igor Bobrowsky's time spent in his CAP interacting with the people changed his mind about Vietnamese civilians, of whom he had not had a favorable opinion while in his prior assignment in the infantry. Serving in CAPs brought a sense of meaning and purpose to the war for Bobrowsky, and he gained respect for the Vietnamese people, realizing they were "human beings, which was quite a change."[133] In a CAP near Phu Bai, Jim Shipp explains that the villagers "went from being 'them' to being people I ate with, fought alongside, and grieved with."[134] In remembering his CAP village, Michael Cousino writes, "I really felt at home there, and I learned to love the Vietnamese people."[135] In December 1965, when the CAP Marines who had spent four months in a village near Phu Bai departed, crying villagers lined the outbound road. A Marine Corps pamphlet attributes bonds between the Americans and South Vietnamese to "the supreme camaraderie of sharing real danger and overcoming it."[136]

More than thirty years after the fall of Saigon in 1975, CAP veterans continue to remember particular villagers whom they had befriended during the war. Former American CAP members curiously and pessimistically wonder what ultimately happened to the villages and their inhabitants. Timothy Duffie was saddened to think of the villagers after he returned to the United States. "I worried about all of my 'family' in Lai Phuoc and Phuoc My. As the years passed, I would often look at the picture Co Hue had given me. That did little more than fan the fires of concern for all of my friends in the village."[137] One of the female villagers from Duffie's CAP village continued to send him letters and pictures after he had departed. Like many veterans of the war, some who served in the program still refuse to return to Vietnam. Former CAP Marine Mike Smith vows that he will never return because "we were sent there to protect those people. I can't face them."[138]

Others have ventured back to Vietnam, in some cases returning to their former CAP villages. Jose Molina returned in 1989 and reunited with some of his CAP villagers, who had erected murals in their homes remembering the Marines.[139] The same year Tom Harvey and eight other former CAP Marines returned to Vietnam. Since that visit, Harvey has reconnected with the PF platoon leader from his CAP.[140]

Although many Americans conformed to life in CAP villages, some failed to adapt. Throughout the program's tenure, CAP commanders removed from the villages Marines who were ultimately deemed detrimental to the overall effectiveness and cohesiveness of the combined unit. Some refused to follow orders while others intentionally disrespected and abused villagers. There also were cases of CAP Marines freezing in combat while on patrol. Numerous historians have attempted to explain the varying behaviors of soldiers in combat. Kyle Longley argues that although fear always accompanied combat soldiers in Vietnam, most were able to overcome it in order to fight effectively for their survival and that of their comrades. But for some, the horrific scenes of combat proved too overwhelming.[141] As Peter Kindsvatter shows, military psychiatrists in Vietnam dealt with soldiers whom they considered "psychiatric casualties" of combat. Some attributed breakdowns during combat to "character and behavior disorders." Yet, as Kindsvatter perceptively observes, combat soldiers knew that every person has his own breaking point. Personality traits may have caused some "psychiatric casualties," but every soldier at some point is susceptible to anxiety and breakdown.[142] The issue then becomes who can overcome the fear and anxiety. Overall, stress, exhaustion, guilt, and fear all influenced the behavior of soldiers, including the Marines and corpsmen in CAPs. With the marginal effectiveness of the program's selection and training processes, some CAP Marines likely brought with them into the villages erratic behavior that had been either overlooked or ignored. After all, many of the Marines who landed in CAP villages had already experienced the volatile and intense emotions of combat. On the other hand, it is possible that some of the combat-hardened Marines ultimately could not adapt mentally to the concept of living with only a handful of their fellow Americans amid an entire Vietnamese village.

The Americans, village civilians, and PF each played critical roles in determining the conduct and behavior of the collective human triumvirate of a CAP village. Faced with numerous perils and uncertainties, the fresh American arrivals to a CAP village had to build a rapport with a civilian population that was experiencing its own trepidations. As the foreign occupants, the Americans had to initiate social interaction in hopes of gradually making the villagers more comfortable with their presence. Through the multitude of civic action projects, most notably MEDCAPs, and daily interaction, the Marines and corpsmen came to understand that creating and sustaining amicable relations with the villagers enhanced military security and their chances of survival. The Americans' transformative perceptions of the villagers, though unintentional, were crucial for bolstering security within the villages. The Marines and corpsmen did not enter CAP villages believing they would forge friendships or empathize with the villagers. After many weeks had passed, the Americans had begun to realize that engaging the civilian population and respecting local customs reaped military benefits, namely, the intelligence offered by the villagers. But by the end of their tours in the program, the Marines and corpsmen did not interact with the civilians solely for military purposes. Many had come to care about the personal safety and well-being of the villagers.

And it was up to the Marines to ensure that safety. For this duty, the Marines had to train the local citizen soldiers, who viewed their military tasks as an obstruction to their daily familial duties in the village. The Marines faced the mission of training the PF, the most marginalized military units in South Vietnam's military.

PF standing in front of the compound headquarters of CAP Delta-1, 1968.
(Courtesy of Tom Harvey)

A group of Marines and PF from CAP Delta-1 on patrol, 1968. (Courtesy of Tom
Harvey)

Village children congregate in front of the school in CAP
3-3-5, 1969. (Courtesy of Tom Harvey)

Two Marines pictured with PF from CAP 1-3-6 in the village of Phu Le, 1968.
(Courtesy of Robert Holm)

Part of CAP 1-3-6 in the village of Phu Le, 1968. (Courtesy of Robert Holm)

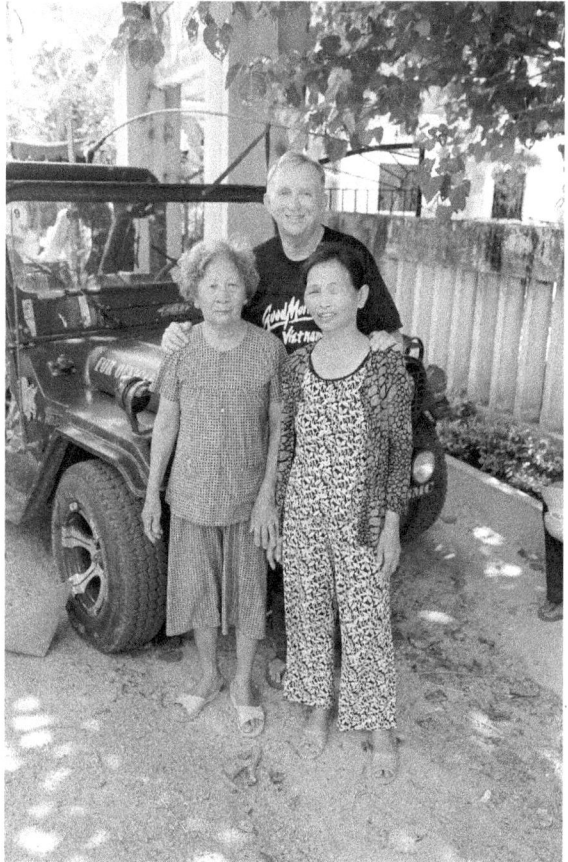

Former CAP commander Robert Holm reunites with two villagers from CAP 1-3-6, 2012. The woman on the left sold soft drinks to the Marines, and the other used to tailor Holm's uniform. (Courtesy of Robert Holm)

Robert Holm of CAP 1-3-6 reunites in 2012 with Hue, who served as the Americans' interpreter in 1968 in the village of Phu Le. (Courtesy of Robert Holm)

PF from CAP 2-7-1, 1970. (Courtesy of Channing Prothro)

PF soldiers stand at attention for inspection, 1968. (Courtesy of Ron Titus)

A group of PF taking a chow break outside the compound of CAP 3-3-5. (Courtesy of Tom Harvey)

CAP Marines and PF open fire on a sniper while on patrol. (VA02088; courtesy of Brig. Gen. Edwin H. Simmons Collection, The Vietnam Center and Archive, Texas Tech University)

CAP 1-3-6 Commander Robert
Holm, 1968. (Courtesy of Robert
Holm)

CAP patrol. (Courtesy of D. Reed)

CHAPTER FIVE

Popular Forces in Combined Action Platoons

It is perhaps only a slight exaggeration to suggest that, on their own, foreign forces cannot defeat an insurgency; the best they can hope for is to create the conditions that will enable local forces to win it for them.

—John A. Nagl

As John Nagl's quote above suggests, the PF ultimately had to win the war themselves. The CAP Marines had to provide the conditions that would allow the local forces to win the war once the United States departed. The program's standard operating procedure assigned the Marines to "motivate, and instill pride, patriotism and aggressiveness in the PF soldier."[1] Yet from the perspective of the Marines in the villages, the PF did not always display the desired qualities as established by the program. After months of training by the Marines, many PF continued to show signs of indifference and a general reluctance to perform their military functions. CAP commander Sgt. Robert Ashe sums up a typical experience for the Marines: "Some days it's like this. The PFs won't go out of the compound. They refuse to go. Period."[2]

According to U.S. Marine and CORDS sources, PF in CAPs collectively became more aggressive and efficient as a fighting force.[3] Testimonies from CAP Marine veterans portray an indigenous force that did improve over time. Yet, the Marines also remember many of the PF as sluggish and incompetent. Although trained to defend their own villages, the PF frequently yielded command authority to the Marines, who gladly took the reins of military control to ensure personal, village, and unit survival. The Americans did not trust the PF enough to give them autonomy of command.

The statistics presented to U.S. Marine commanders showed much improvement in the PF. However, many PF did not try independently to improve their combat effectiveness; the Marines' aggressiveness forced them to participate in their basic military duties. When they left assigned villages, very few CAP Marines had confidence that the PF would continue to operate as they had during the American presence.[4] The PF did not join the territorial forces out of *rage militaire* or because of a patriotic or ideological drive to defend their country, as many Americans had. PF volunteered for the territorial forces to avoid being conscripted into a regular ARVN unit, duty that entailed leaving their homes and families for an extended period.[5] The PF were foremost farmers and fishermen—and soldiers on the side. As the 1969 MACV handbook on territorial forces warned U.S. advisors, a member of the PF was not particularly a "hard charger."[6] The training and general military knowledge of the Marines immeasurably eclipsed that of the PF. Thus, considering the small number of Marines and the low motivational state of PF in the villages, if the Americans were indifferent, the situation could quickly turn to disaster for CAP units.

Westmoreland stated that Americans could help provide security and civic action in the villages, but only the South Vietnamese government could ultimately achieve success with pacification.[7] Historian Mark Moyar has argued that CAPs provided a quick fix for improving the PF, but in the long term, the hands-on training allowed the South Vietnamese to stall in their efforts to improve their military leadership problems.[8] In other words, with the Marines leading the way in daily military operations, the PF let them do most of the work. In retrospect, these are valid critiques of CAPs. According to historian Andrew Wiest, the U.S. military in general tried to win the war for the South Vietnamese.[9] A microcosm for the entire U.S. military effort in Vietnam, CAP Marines attempted to win the war for the PF and the civilians in their villages. From the perspective of the Marine NCOs and enlisted personnel in CAP villages, PF represented the best and worst of the South Vietnamese military. Some Marines dealt with effective, trustworthy PF, while others experienced ineptitude and ineffectiveness from their South Vietnamese counterparts.

When one examines the history of the PF, the indifference and ineffectiveness that characterized many of them comes as no surprise. The PF were a force neglected by both the U.S. military and the South Vietnamese government. The territorial forces of South Vietnam consisted of the RF and PF, dedicated to local security and pacification at the district and village levels, respectively. Throughout the Vietnam War, the territorial forces suffered from low pay, high desertion rates (with the exception of those in CAPs), and a lack of training and motivation. The PF, operating in the villages where many of them resided, occupied and defended critical ground in the war. The VC thrived off the rural population the PF protected. Compared to South Vietnamese cities, under firm RVN-U.S. control, the villages and their occupants were more accessible to the rural-based VC, offering an abundance of potential recruits, allies, food, and hideouts. With the VC constantly seeking to gain political and military momentum in the villages, the territorial forces of South Vietnam suffered high casualty rates. Except for 1968, when the Tet Offensive took the war to the urban areas of South Vietnam, the territorial forces had a higher combat death rate than any other segment of the U.S. or South Vietnamese armies.[10] Although the PF protected critical ground in the war, the RVN and U.S. military alike marginalized the territorial forces. By 1966, just one year after the war's inception, the scant attention given to the territorial forces convinced one American study group to brand them "the stepchildren of the GVN."[11] The territorial forces obviously needed help, a truth that was obvious even before the arrival of U.S. combat troops in 1965. Yet until the creation of mobile advisory teams and mobile training teams after the 1968 Tet Offensive, CAPs constituted the only American military endeavor solely dedicated to improving the PF. The RVN attempted to use South Vietnamese personnel to improve the territorial forces. In many cases, however, the South Vietnamese teams sent to the villages were just as ineffective and unmotivated as the PF.

When the U.S. Marines landed in 1965, the combined U.S.-RVN military apparatus had not prepared the PF to fight the upcoming war effectively by themselves. After the Geneva Accords in 1954 that created the Republic of Vietnam, South Vietnam split its military into regular and territorial forces. Fixated on attrition-based warfare similar to World War II and Korea, the U.S. military advisory mission, the Military Assistance and Advisory Group opted to leave the territorial forces out of its military assistance program, which fostered poor leadership and lackluster training among the PF ranks. MAAG prepared the armed forces of South Vietnam for a conventional invasion from Communist forces to the north, similar to North Korea's assault into South Korea in June 1950. Gen. Sam Williams, the MAAG commander in 1960, asserted that "if we as Americans, either alone or as part of SEATO, become engaged in a hot war in [Southeast Asia,] it will be against Chinese forces primarily and not against the Viet Cong or North Vietnam."[12] Preparations for a conventional conflict rendered the territorial forces dedicated to pacification less relevant in the RVN's and U.S. military's grand strategy. In the late 1950s, the ineffectiveness of the territorial forces, then called the Civil Guard and Self-Defense Corps, forced MAAG to dispatch regular ARVN units to the countryside to assist in pacifying the villages—the same ARVN personnel that had trained as a conventional force under U.S. guidance.[13] ARVN units assigned for rural pacification performed their duties reluctantly, as most of them wanted to conduct larger mobile operations outside the villages. In 1964, when insurgents in rural areas began to pose problems for the U.S. military command structure, South Vietnam's Ministry of Defense assumed control of the territorial forces, giving the RF and PF full status as members of the Republic of Vietnam's Armed Forces (RVNAF). By the time U.S. combat forces arrived in 1965, the territorial forces had improved over the previous decade but, according to Andrew Wiest, "the attention and improvement had come ten years too late and could not quickly overcome years of neglect and stagnation."[14]

After the assassination of Ngo Dinh Diem in 1963, the VC increased its presence in South Vietnamese villages. To counter the insurgent threat, the South Vietnamese government enacted several programs to gain control of the villages. In 1965, Westmoreland introduced the Motivation Indoctrination Program to help motivate territorial forces. Until 1967, the program remained a staple of bolstering the PF's awareness of their association with the RVN's pacification plans. Yet by 1967, only 50 percent of the territorial forces had encountered indoctrination teams in their villages. In the spring of 1967, revolutionary development (RD) became the catchphrase for pacification in South Vietnam. With the reduction of the Motivation Indoctrination Program in 1968, the RVN had increasingly relied on RD teams to persuade villagers to support South Vietnam. Under this program, fifty-nine-man RD teams entered villages to recruit for the armed forces, help with community projects, and politically indoc-trinate villagers. Plagued with poor leadership and high desertion rates, many RD teams failed to accomplish their tasks.[15] Theoretically, the PF were supposed to help the RDs provide security. However, RD cadres treated PF as their inferiors. Moreover, the cadres failed to promote the legitimacy of the national government in the villages, which stems in large part from the failure of the urban-based RD teams to establish amicable relations with the villagers.[16] During Vietnamization, CORDS represen-tatives in I Corps continued to notice an inefficient RD force.[17] In 1970, South Vietnamese in contact with RDs complained about inadequate se-curity and a general lack of coordination with the territorial forces.[18] Few RDs had lived in the villages in which they traveled, and when the South Vietnamese pacification teams made contact with CAPs, the Marines noted the negative impact they had on the people. When not hiding out from suspicious noises at night or drinking beer during the day, RDs in one CAP village tried to "coax the girls into the bushes." Bing West recalls that when the RDs departed the CAP village of Binh Nghia, they had failed to convince the villagers to despise the VC.[19] In 1968, RD teams in Quang Nam were notorious for leaving their assigned villages just before nightfall to seek refuge for the night in CAP villages.[20]

To measure the effectiveness of PF, MACV developed a territorial forces evaluation system (TFES) that required district senior advisors to submit monthly reports on PF platoons within their districts. In 1968, the Fourth CAG commander noted that the TFES would help measure the success of CAPs, but he also recognized the scheme had several weak-nesses because there were not enough district advisors for the numerous PF platoons within each district. Inaccuracy saturated the TFES: some of the numbers of active PF in certain CAPs as reported by the district failed to match the figures given by the Marines. In Quang Nam, four different CAPs reported lower PF numbers than the district headquarters did on its TFES ledgers. In the most extreme cases, one Marine unit reported eighteen PF available for duty, while the TFES figure stood at thirty-one. During the same period, one CAP in the I Corps province of Thua Thien reported as few as six PF, while the TFES report stated that the unit con-tained twenty-eight. The problem also occurred in the southernmost I Corps province Quang Ngai, where seven CAPs recorded twenty or fewer PF, while the TFES numbers were higher.[21] District representatives who failed to inspect all the PF within their sector often resorted to guessing or averaging the figures from villages they did visit.

The South Vietnamese officers left over from the French-controlled Vietnamese National Army brought politicization with them into the U.S.-backed ARVN.[22] In a corrupt South Vietnamese government and military, with officer appointments and promotions based on political loyalties and financial connections rather than merit, the PF had little chance for advancement. To qualify for South Vietnamese officer candidate school, one needed to hold a baccalaureate degree, awarded after twelve years of expensive formal schooling in the cities; such was financially and logistically out of reach for PF. The PF had no system for promotion, and the only incentive to serve, besides an inadequate paycheck, came in the form of military decorations. The South Vietnamese Joint General Staff had supported the creation of a system of rank and method of promotion for the PF, but by 1970 that idea had dissolved.[23] According to a 1968 report on PF from the Center for International Studies, "Being a platoon leader is a dead end."[24]

South Vietnamese political officials' corruption played a large role in depleting the morale of PF. Province chiefs were high-ranking ARVN officers appointed by the South Vietnamese corps commander, usually a general-grade officer in the RVNAF. Each province chief controlled the purse strings and the dispersing of material to RVNAF forces within his jurisdiction. The province leaders were notorious for filling district chief positions with persons who paid their superiors for the important political position. Province chiefs, overburdened by military and civilian responsibilities, were frequently forced to delegate their decision-making and administrative duties to the various district headquarters. This gave district chiefs the power to appoint their political cronies to village chief positions.[25] In the early years of the program, CAPs routinely begged district chiefs to add more PF to an already undermanned ten- or twelve-person platoon. To prevent the district chief from removing CAP PF whom the Marines deemed effective, one of the Americans would have to plead to keep the South Vietnamese soldiers in the village. Some district chiefs required payments from Marine commanders in return for attaching PF to CAP units. To comply with one such district chief's demands, William Corson solicited money from fellow Marines to purchase PF for CAPs in need of more troops.[26]

Village chiefs, who received and distributed the PF's monthly payments, often stole the money, keeping some for their own personal use and distributing the rest to district headquarters for job security. Village chiefs also provided their superiors with rosters of "ghost soldiers," a list of nonexistent PF serving in the hamlets. Since villagers had to pay taxes for the PF, a larger platoon warranted more of the civilians' money, which first landed in the hands of the village chiefs. Thus, "ghost soldiers" became a common way for rampantly corrupt village chiefs to pad their pocketbooks.[27] In one I Corps district, ten months passed before the PF received their guaranteed monthly allotments. After an investigation, U.S. military officials found that the district chief had used the money to pay his own office expenses.[28] In 1969, the South Vietnamese government allowed villagers to elect their own chiefs. Also during Vietnamization, PF fell under the operational control of the village chiefs. Yet, knowing that communicating with Americans and PF made them a target for assassination attempts by the VC, village chiefs rarely stayed in the villages. In 1971, under the direction of William Colby, CORDS argued that the decision to give village chiefs more control was "key" to "the improvement of PF effectiveness."[29] CORDS' positive report in 1971 of territorial forces speaks to the disconnection between the high-ranking pacification analysts and the performance of the PF at the local level. The optimistic analysis of South Vietnamese territorial forces applauds village chiefs for molding the PF into a highly motivated "superior form of full-time militia." However, the document fails to address the consequences of corruption in the villages. Nor does it consider how absent village chiefs impacted the performance, motivation, and morale of the PF.

Isolated from the PF in the villages, each district headquarters held responsibility for monitoring the security situation in its zone of responsibility. However, district officials rarely made an active effort to improve morale or show the least concern about the welfare of the PF. Without an efficient or consistent monitoring system from district headquarters, PF platoon leaders could neglect training and motivating their soldiers and the district chiefs would never know.

One of the consequences of the disjointed, corrupt South Vietnamese government was that its military was never a unified ideological force, as was its North Vietnamese counterpart.[30] In the words of historian Alexander Woodside, the South Vietnamese army did not satisfy "nationalist aspirations."[31] This lack of unity affected all levels of the South Vietnamese military, especially the PF. A 1967 Simulmatics Corporation report concluded that the PF lacked an understanding of South Vietnam's long-term political goals. The villagers and PF across South Vietnam had "little feeling of national identification and unity." Of the more than one thousand subjects (forty-five RF squads and forty-seven PF squads) interviewed, only one-third of the territorial forces said they "liked" the RVN government. The interviewees hoped to achieve peace rather than victory for South Vietnam. According to the report, the PF's "commitment to ideological and national objectives in the war against the VC is highly questionable." Local and family defense functioned as the primary motivating factor for the PF, even if that meant accommodation with the VC. Some PF did show a disdain for the VC and foresaw a gloomy future if it won the war. However, PF feared a VC victory on a personal, not an ideological, level. As the report argued, "The typical PF acted just like the peasant he was on the day he joined the PF."[32]

The training procedure for territorial forces was afflicted with inadequate facilities, ineffective instructors, and a general lack of discipline. Overcrowded lecture rooms and apathetic instructors weakened the training process from the moment a recruit entered the facility.[33] Standard PF instruction consisted of five weeks of basic combat training and an additional four weeks of advanced individual training. Once the men had landed in a platoon, newly activated PF units were supposed to receive another three weeks of basic unit training before arriving in their assigned area of operations. However, most province chiefs proved hesitant to relinquish the primary source of village security.[34] In November 1969, MACV commander Creighton Abrams vented his frustration with the ineffective training centers during his weekly meetings with top-ranking military and civilian officials: "These PF platoons are coming out of those goddamn training centers, and they couldn't fight their way of a paper sack."[35]

Perhaps the most influential morale disincentive for the PF was their low pay, about half that of the regular ARVN soldier. The South Vietnamese government justified this by contending that territorial forces served near their homes, unlike the regular ARVN soldiers, who spent most, if not all, their time in the war away from their families. The corruption that infested every level of the South Vietnamese provincial chain of command added to the frustrating financial situation for the PF. In late

1966, Krulak argued that the low pay of PF undermined the potential for the program's growth, crediting it with the declining numbers of PF in I Corps. Krulak, perhaps the most vocal champion for CAPs, was concerned that many villagers were bypassing the opportunity to join the PF and instead volunteering for safer assignments with the South Vietnamese government or military. Most notably in the Da Nang area, potential PF recruits opted for civilian jobs in the city, where they enjoyed relative safety and higher pay.[36] The low numbers of PF volunteers in 1966, coupled with their high casualty rate, made improving the territorial forces a necessity for IIIMAF and the program. From the program's perspective, an increase in the number of CAPs would result in visibly heightened security for the villagers, which in turn could possibly persuade more civilians to join the PF.

By 1966, PF in I Corps had already begun to sustain substantially high casualty rates. In the month of February, although PF comprised only 20 percent of all South Vietnamese military personnel in I Corps, 54 percent of all RVNAF soldiers killed came from their ranks.[37] Their poor training and inexperience undoubtedly played a role in the disproportionate casualty rate. In addition, the PF were popular targets for the VC seeking to take control of the villages. The high PF casualty rate in I Corps alarmed Marine leaders. Krulak noted the increasing probability that I Corps villages would fall to VC control at an irretrievably rapid rate if the Marines did not intervene. As IIIMAF and FMFPAC headquarters continued to receive positive reports from the initial combined Marine-PF units, expanding the number of CAPs fulfilled Walt's and Krulak's hopes of decreasing the number of PF casualties while simultaneously bolstering their effectiveness. The PF were crucial pieces in pacification, but it became obvious that the village militias alone could not successfully secure their areas of operation. From the perspective of Marine commanders, to keep, gain, or regain control of I Corps villages with PF platoons, Marines needed to train and stay with the local militias.

When arriving in a newly created CAP village, the Marines quickly recognized that the PF needed massive upgrades in virtually every facet of military training. Years of neglect from the RVN and U.S. military had presented CAP Marines with the seemingly insurmountable task of turning the PF into a formidable fighting force. In the village of Binh Nghia, CAP Marines watched dumbfounded during target practice as PF missed a five-foot-high target fifty yards away.[38] Their inaccuracy was no surprise, however, considering the PF were products of provincial training centers that carried out live-fire exercises with the entire class firing their rifles simultaneously at the same target.[39] In addition to education on their weapons, PF required refresher instruction on small-unit tactics, which the provincial training centers had neglected.[40] Improving the effectiveness of the PF in CAPs rested squarely on the shoulders of the Marines.

The Marines trained the PF daily with instruction on tactics, ambush techniques, proper employment for claymore mines, and how to clean their weapons. On-the-job training was crucial in the villages. The PF's confidence increased from watching the Marines and mimicking their actions on patrols, especially when the Marines stood their ground during a firefight. In some cases, the PF actively participated in the training process. However, many showed their lack of interest in training and patrolling, partly because they preferred to tend to family duties during the day. In one CAP, the daily call to prepare for upcoming patrols brought only one PF to the rendezvous point.[41]

The U.S.-RVN military structure of dual command did not allow American forces to command Vietnamese troops. In CAPs, the Marine commander was supposed to retain exclusive control of his American forces, leaving the PF platoon leader in sole command of the South Vietnamese. Americans and PF often collaborated on training and administrative duties and in planning ambush and patrol locations. During a firefight, however, the PF usually yielded command and control to the more experienced and better trained U.S. Marines. Yet the PF did not always participate in firefights. In the event that an enemy force approaching the village outnumbered the Marines, immediate combat support for the Americans rested with an inexperienced and generally lethargic PF force.

The program demanded that PF always accompany Marines on patrol, but the militia forces frequently shrank from aggressive activity that might spark a firefight with the VC. At night, when the VC was most active, PF often avoided patrols. Many CAP veterans agree with the conclusion of John Akins, who contends in his memoir *Nam Au Go Go* that the PF "survived by avoiding trouble, and we were looking for trouble."[42] A 1969 report on the program found that "in more cases than it is polite to recount, when the VC come, the PF hides."[43] After several instances of PF running from danger in one CAP, the Americans began calling them "PF

Flyers" after the American-made high-top sneakers. The corpsman from that CAP writes, "We started calling them PF Flyers because when things got tough, they flew the coop."[44] PF tended to village duties during the day and were disinclined to participate in combined military activity until the evening hours.[45] Thus, the PF did not have ample time to rehearse and discuss night activities before they commenced. PF expected the Marines to leave them alone during the day, especially if they had participated in a patrol the previous night. With the arrival of the CAP Marines, many of the older PF who had served in the military for a couple decades quite naturally disliked taking orders from nineteen-year-old foreigners.[46] PF moved quickly on patrols, forcing the Marines to move from a methodical march to a louder trot, enhancing the possibility of detection from the VC. The brisk pace of the patrols cracked branches, and whispered orders from Marines turned into frustrated shouts as the PF gradually distanced themselves from the predetermined route. In some cases, PF deliberately made as much noise as possible to signal their presence to any VC hidden in the jungle in hopes of averting combat. The Marines found that the PF were able to detect sights and sounds unnoticeable to the Americans. Barry Goodson recalls an instance when a PF saved an entire CAP patrol by identifying several hidden booby traps that the Americans themselves likely would not have detected. For example, the PF spotted a vine on the jungle floor that acted as a trip wire for grenades hanging from tree branches. Goodson believes that without the observant PF, someone in the patrol likely would have activated the booby trap.[47] Yet when a firefight ensued, the Marines did most of the fighting. One Marine explained, "It is difficult to get the PFs to open fire on the VC. So we use the PFs as our eyes and ears. It is the Marines who do the actual fighting. You cannot always depend on the PFs to advance with the Marines."[48]

Some Marines wholeheartedly believed that certain PF were actually VC.[49] Program veterans today still wonder if they wasted their time in CAPs training the enemy. Several incidents ignited the Marines' suspicion. Disconnected wires the Marines had attached to M-18A1 claymore mines were one of the more frequent warning signs. The Marines and PF strategically placed the mines outside the village perimeters. Claymores were detonated by a remote device, and in CAPs the handheld detonator remained within reach of the Marines twenty-four hours a day. Yet the CAP Marines could not monitor all areas of the village at any given time.

Although at least a few Marines always remained on watch, the separation between the individual hamlets that comprised a village created plenty of open space for people to maneuver undetected. Villagers, including PF, could easily roam unnoticed by the Marines to particular spots inside and outside the perimeter, especially if those individuals knew the exact schedule of the Americans' daily operations. Since the Marines and the PF knew the locations of all the hidden mines, from the American perspective, the culprits guilty of the disabled mines had to come from the CAP unit. Every Marine knew the location of his fellow Americans at all times, allowing them to narrow the possible suspects down to PF. Of course, the problem with confronting the PF was that if they had cut the wires or informed the VC of the claymores' locations, they would never admit their guilt to the Marines. Although in these cases the Americans could only suspect that PF had disabled the mines, their trust of the South Vietnamese soldiers disappeared forever. There were, however, instances in which Marine suspicions that PF supported or fought alongside the VC were confirmed. Marines in one village caught two PF hidden in a tunnel planning an attack on the CAP unit. On numerous occasions, Marines found PF among the deceased enemy combatants after a firefight.[50]

In August 1967, in the village of Hoa Hiep, villagers gathered to watch a presentation of the film *John F. Kennedy: Years of Lightning, Day of Drums,* complete with Vietnamese narration provided by a South Vietnamese psychological warfare team. With the villagers' and Marines' attention on the film, the VC had secretly infiltrated the village and booby-trapped the psychological warfare team's truck. A CAP Marine wandering the village found the grenade by chance and disarmed it. When the Marine informed his CAP commander, Sgt. Gary Smith, of the incident, they both realized that three PF had failed to report for duty that night. One of those PF had, curiously, escaped unscathed from a major firefight with the VC the previous fall. As the Marines congregated to discuss the incident, they recalled other activities the PF had engaged in that in hindsight seemed suspicious. Sgt. Smith exclaimed, "You remember a couple months ago we caught him signaling one night. I thought he was just fooling around with the light. Now I don't know. It didn't bother me till I heard about the grenade." Although they were not able to accumulate enough evidence to accuse the suspected PF of supporting the VC, the incident of the booby-trapped truck made the Marines uneasy about conducting military operations with the PF. With distrust between the Americans and PF on the rise, then corporal Ron Schaedel had duty in a bunker one night with a PF in Hoa Hiep. Every Marine on bunker duty always had a PF at his side. They would take turns sleeping and standing guard, but Schaedel never fell asleep that night out of concerns about the PF's loyalty. "Man, I just don't know whose side that guy is on," Schaedel said. "I don't want to go to sleep in that bunker."[51]

American frustrations with some of the PF's lack of motivation spawned physical confrontations. Even friendly volleyball matches between the Americans and PF could escalate into fistfights.[52] The Americans' aggravations stemmed in part from the fact that according to the program's standard operating procedure, the Marines could not give the PF military orders of any kind. In 1969, in the CAP village of Kim Lien in Quang Nam province, the PF accused a Marine corporal leading a patrol of intentionally shooting and killing one of their comrades.[53] In the days preceding the incident, the Marine had allegedly threatened to kill members of the local forces for their lackadaisical behavior on patrols. Col. Danowitz, the assistant chief of staff of the program at the time, noted the tension between the Americans and PF when he came to the village to investigate the incident. Danowitz heard from the Marines that on certain nights, only twelve of the forty available PF reported for duty. Moreover, out of the twelve, only three agreed to patrol with the Marines. Realizing how perilous the situation was, Danowitz updated Second CAG of the situation, ordering headquarters to continue to monitor the Marines from the village if problems persisted.[54]

Despite deficiencies in the PF, the Marines learned more about the intricacies of jungle warfare from them than the U.S. military could ever have provided. As Bing West recalls about a PF, he could "worm his way thought a dry thicket without breaking a twig and he could spot a Viet Cong on nights so dark other patrollers could not see the man in front of them."[55] CAP Marine Chuck Ratliff recalls an occasion when a PF saved his life during a night patrol: "As I was looking back at the rest of the patrol, I noticed that one of the PF seemed to be aiming his rifle right at my head. As he fired, I heard the bullet whiz past my head. As the round went by I looked the other way. I saw a VC soldier dressed in black falling to the ground with a bullet in his forehead. I only saw him for a few seconds but to this day I can still picture his face in my mind.

The PF had seen this VC sneak up on me and didn't have time to warn me, so he shot him."[56]

Accustomed to patrolling with much larger units, the Marines noisily stomped through the jungles with their heavy boots. The PF demonstrated that walking with sneakers, PF-issued rubber-soled boots, or no shoes at all limited the noise. As a Marine's time in a CAP progressed, he commonly discarded elements of his uniform, even patrolling without helmets, shirts, or shoes.[57] The program associated the Marines' appearance with a lack of discipline and worried that it might result in low morale or indifference. The Marines' habit of jettisoning parts of their standard-issue uniforms, their "five o'clock shadows," and their relatively longer hair did not meet the traditional grooming standards of the Marine Corps. But the Marines in the villages were simply adapting to the environment, the climate, and the people. From the perspective of Bing West, the Americans in Binh Nghia "were beginning to feel at home in the village, with its guerrillas and PF, fishermen and farmers, women and children. Many of the Marines let months go by without writing a letter or reading a newspaper. The radius of their world was two miles."[58]

PF possessed an unmatched knowledge of the local terrain. They knew the best shortcuts and hideouts outside the village, enabling the CAP unit to maneuver more effectively while on patrol. Moreover, the PF interacted with the villagers more frequently than the Americans did. Without a language barrier, the PF could communicate more efficiently with the villagers to gain useful intelligence. With generations of families living in the villages, the PF had local knowledge the Marines could never hope to attain; they knew who lived in what house and which family might have relatives supporting the VC. One CAP Marine recalls that the PF in his unit had familial connections with 160 out of the 200 families living in the village.[59] PF could easily detect the potential danger in movement and sounds that seemed normal to the Americans. Some of the older PF had lived in their village for decades, making them highly sensitive to anything out of the ordinary. Thus, PF could alert the Marines to possible pending attacks and notify the Americans of VC infiltrators in the village. Civilians frequently divulged intelligence to the PF, who then passed the information to the Marine CAP commander. For example, Bing West shows that the PF in Binh Nghia were informed by the villagers when the VC began forming new trails outside the village to avoid contact with the

CAPs.[60] When the VC did successfully infiltrate a CAP village, usually at night, they recruited relatives of the PF. In some cases, the PF fought their own family members while on patrol with the Marines. After a firefight near the CAP village of Tuy Loan, two of the resulting deaths were brothers, one PF and the other VC.[61]

When PF proved themselves in a battle that protected the villagers from harm, it resulted in increased rapport among all elements of a CAP village. In a chapter entitled "Acceptance" from *The Village,* Bing West describes how the villagers of Binh Nghia revered the PF after the combined unit had repulsed an NVA battalion-sized attack.[62] Although training the PF was a frustrating experience for the Americans, many Marines trusted and respected their counterparts.[63] In a CAP village in Quang Tri, Marines were devastated to learn of the death of one of the most respected PF in their unit. The Marines gave the PF a twenty-one-gun salute before his family buried his remains. One of the Marines from that CAP recalls, "The tears we shed for this dear friend were as real as they were for any of the American friends we lost during the war."[64]

Today, it would be difficult to find a veteran of the program who departed his village confident that the PF would continue to operate effectively without the American presence.[65] Most of the PF platoons that stayed in villages after the relocation of the Marines backpedaled to their previous ways of inconsistency, indifference, and lethargy.[66] Later in the war, when VC activity had diminished in certain CAP villages, the PF platoons patrolled their area of operations without the Marines.[67] In some cases, the PF continued to provide effective resistance to the VC. After seventeen months in Binh Nghia, the Americans departed the village, leaving the military duties to the PF. Bing West argues that after the Marines left, the PF patrolled "like cops on a beat."[68] After two months of patrolling without Marine assistance, the PF of Binh Nghia held their own, encountering no serious incidents with the VC. When a mixed NVA-VC unit attacked Binh Nghia, the PF, via their intelligence network, ambushed the oncoming enemy patrol and successfully repulsed the attack. By 1970, the village of Binh Nghia had been so peaceful that the American district advisor declared the village an R and R center. However, some instances of PF choosing to patrol without the Marines had disastrous results. In January 1969, a CAP PF leader in the province of Quang Tri, with the Marines still present, insisted on executing an ambush without American help. After the Marines approved the proposed ambush plans, the PF left the compound and ultimately engaged the VC in battle. Ten PF died.

By the end of 1968, of the thirty-two villages where CAPs had dissolved, the majority of the PF platoons discontinued civic action, made very little contact with the enemy, and suffered from severe supply shortages. One must also note that many of the former CAP villages housed replacement PF units. Moreover, when the CAP Marines left, nearby American artillery and air components were reluctant to provide support for the PF because of their insecure radio networks. The Americans feared that the Vietnamese person calling for artillery or air support could be VC.[69]

During Vietnamization, the United States and RVN broadened the PF's target areas of operation beyond their home villages. The 1970 RVN pacification and development plan called for the PF to expand in size and to begin replacing RF companies, ostensibly giving the PF a greater role in territorial security. In 1971, Operation Lam Son 719 tested the ARVN's fighting capabilities without the assistance of U.S. combat troops. The operation, which sent the First ARVN Division into southeast Laos, neighboring Quang Tri province in I Corps, did not reap the benefits the RVN, the ARVN, and U.S. military advisors had anticipated. It seemed clear that the low morale of ARVN troops returning from the operation had spread throughout the entire South Vietnamese population.[70]

By 1972, 250,000 PF soldiers provided district and village protection for the oncoming NVA Easter Offensive later that spring. At this point of the war, with only U.S. military advisors available, local ARVN divisions within each province had orders to occupy any territory lost by the territorial forces. In Thua Thien, the movement of First ARVN Division units to counter the NVA threat forced the territorial forces to cover a larger area of operations to make for the territory formerly occupied by the regular South Vietnamese soldiers.[71] During the Easter Offensive, although the territorial forces in Quang Tri had "disintegrated" as a fighting force, the PF in the remaining I Corps provinces held their ground.[72]

By the end of March 1975, the NVA, en route to Saigon, had taken control of most of I Corps and had captured more than one hundred thousand South Vietnamese soldiers in the process.[73] During the RVNAF's retreat from I Corps, the RVNAF "lost" two infantry divisions, four Vietnamese Ranger groups, three armored cavalry squadrons, twenty-seven

RF battalions, and more than twelve hundred PF.[74] On 30 April 1975, the Communist forces from North Vietnam ended the war by sacking Saigon, unifying the country as the new Socialist Republic of Vietnam. The South Vietnamese military and political officers and officials who either chose to remain in Vietnam or failed to leave before the Communist takeover would endure "reeducation" under the new regime. It is nearly impossible to find out exactly what happened to the PF soldiers in CAPs who, like thousands of their countrymen, could not elude a Communist "reeducation" camp. The archives in Vietnam have yet to expose all of the available material relating to U.S. involvement in Southeast Asia.

Simply put, the Marines in CAPs were far better prepared than the PF for military tasks in the villages. The well-trained riflemen in the Marine Corps stood in stark contrast militarily with the villagers who doubled as PF. The latter had to tend to domestic duties and try to ensure the health and safety of their families. The Americans fought the war thousands of miles from their homes; the PF fought the war in their homes. While CAPs did have their successful elements, the handling of the PF represents one of the program's shortcomings. This is not to argue that program leaders or the Marines in the villages did not try to rectify the poor discipline and lack of motivation among the PF ranks. After all, the CAP Marines often took control and enforced the required military duties. Yet this gets to the heart of the issue: without a full commitment from the PF, the CAP Marines generally had to fight for them, not always with them.

The Marines in the villages encountered numerous difficulties in dealing with the PF. From the American perspective, an overall lack of trust was a serious issue. Further, most Americans in CAP villages had their fair share of personal squabbles with PF who refused to listen to the Marines. While the Americans fighting in the villages attempted to overcome military and cultural obstacles, the colonels and generals of the Marine Corps in Vietnam had their own challenges to confront—and their obstacles came from their military counterparts in the U.S. Army.

CHAPTER SIX

The Combined Action Program and U.S. Military Strategy in Vietnam

I think the Army people thought of the CAP as something the Marines had started, so that was why Saigon didn't adopt it nationwide.
—William Westmoreland

During the war, Westmoreland told the FMFPAC commander Victor "Brute" Krulak that fighting with CAPs "will take too long." The quick-witted Krulak, the most vocal Marine opponent of the U.S. Army during the war, quickly responded, "Your way will take forever."[1] This brief but direct exchange between Westmoreland and Krulak epitomized the blame game that took place between the U.S. Army and Marine Corps over which service was applying the correct strategy in Vietnam. Caught in the middle of the interservice debate was the Combined Action Program.

CAPs undermined the army's institutional obsession with conventional war, which Andrew Krepinevich has termed "the Concept."[2] For four years beginning in June 1964, Gen. William C. Westmoreland served as the commander of MACV, the unified command structure for the U.S. military in Vietnam. Westmoreland had operational control over all U.S. forces in South Vietnam. From his headquarters in Saigon, the MACV commander formulated and finalized all major strategic decisions for the U.S. military. During his four years of command in Vietnam, Westmoreland continued to enforce the attrition-based strategy that his predecessor and inaugural MACV commander, Gen. Paul Harkins, had applied during his two-year stint in Saigon. Westmoreland used U.S. mobility superior firepower, and advanced technology to attain high enemy body counts. U.S. forces scoured the jungles of South Vietnam to search for and destroy enemy main force units, and these missions were the primary means of measuring success (enemy bodies) in the war of attrition.[3] The MACV commander held steadfast to the belief that to win the war, the U.S. military had to apply overwhelming offensive force, as it had in all the major wars of the twentieth century.

During his time as IIIMAF commander, Lt. Gen. Lewis Walt gradually began to realize that a strategy determined by the number of dead enemy bodies was not a wise choice in Vietnam. As IIIMAF commander from June 1965 to June 1967, Walt had operational control of the U.S. Marines in I Corps. Although he came from an institution that had its own historical attachment to conventional war, Walt believed in and implemented a strategy that differed from Westmoreland's. Unlike Westmoreland, Walt defined a successful body count as "three thousand healthy, secure people, living decently, with hope for the future."[4] Walt, as well as his successors at IIIMAF headquarters, pursued a strategy that paid ample attention to aiding and securing the rural population. Yet this contradicted William Westmoreland's insistence on seizing the offensive initiative via search and destroy missions.

Confining squads of Marines to areas of operation averaging two square miles did not fit the strategic mold Westmoreland had formed as MACV commander. Living in the villages, interacting with the people, and training the PF did not require an overwhelming use of firepower and superior technology. CAP Marines accessed artillery and air support when needed, but they rarely participated in large offensive airmobile assaults or "search and destroy" missions outside their small areas of operation.

When the Marines landed in 1965, they had already mapped plans to identify heavily populated urban areas along the I Corps coast; they would establish a secure perimeter around each of those enclaves. As the war progressed, each enclave would expand like an "inkblot" once more manpower became available to provide security to the people within each area of responsibility. Ultimately, the Marines envisioned one large "inkblot" that would blanket a significant portion of I Corps. Westmoreland hated this strategy because it inherently focused on the people and institutions within the enclaves. The MACV commander wanted the Marines to break out of the enclaves to find the enemy's main force units hidden in the jungles of I Corps.

Although the Marine enclaves contradicted Westmoreland and MACV's strategy, the war of attrition played only a minor role in the program's failure to escalate at the rate CAP proponents envisioned. The program hoped to have 114 CAPs in I Corps by the end of 1967, a number not reached until two years later. Lt. Col. William Corson, who served as director of the program in 1967, dreamed of having one CAP for every two hamlets in I Corps, warranting a force of more than ten thousand Marines and thirty thousand PF.[5] Numerous program veterans have identified the army, MACV, and CORDS as jealous institutions that posed the greatest hindrance to the success of CAPs.[6]

Some civilians and military officers who dealt directly with CAPs did not specifically blame the army for its strategy or for its generally pessimistic assessments of the program. John Tolnay, Second CAG commander during the deactivation phase of the program, argued that to blame only Westmoreland for the program's demise is an "oversimplification," and that the MACV commander "was not the sole reason for not expanding [CAPs]."[7] Tom Collier, who served as the Thua Thien province chief in I Corps, writes, "There was a lot of controversy over tactics and strategies in Vietnam, and CAP was caught in a storm of conflicting views: not just a simple Army v. Marine rivalry."[8]

The general U.S. manpower shortage in Vietnam provided the most formidable barrier to program growth. During the first three years of the war, Marines participated in numerous battalion-sized or larger operations in I Corps while using a portion of IIIMAF resources for pacification efforts. At the peak of the war, I Corps housed seventy-five thousand Marines, five thousand of whom resided in the program. IIIMAF could not afford to commit all of its resources to either the war of attrition or pacification. The Marines did see value in large-unit operations that prevented enemy main force units from disturbing the major urban centers of I Corps. In 1966 and 1967, when the NVA began increasing its attacks into northern I Corps, IIIMAF matched the enemy's presence by concentrating its own main line units near the DMZ. Realizing the dire situation near the DMZ but also staying true to their efforts to provide for the rural villages, Marines had to balance their use of main force units with pacification. "The other war" held high value in the minds of the Marines because pacification hindered the ability of the VC to infiltrate the rural villages that offered food, hideouts, and potential recruits. Yet without the luxury of unlimited manpower, to ignore one war would prove disastrous for the other. The reasons for what program leaders deemed a slow growth rate in CAPs did not stem from a systematic attempt by the army to eliminate the Marines' "other war." Lack of manpower was the main culprit. The Marines, however, rarely considered this when accusing people and institutions of trying to undermine their strategy.

While Westmoreland did not completely overlook the importance of pacification, he was the foremost champion of the belief that to gain any ground with counterinsurgency and pacification, the war of attrition must first reap success. In Westmoreland's words, a commander "wins no battles by sitting back waiting for the enemy to come to him."[9] As a result, Neil Sheehan argues, pacification "shriveled" under Westmoreland's command.[10] In Douglas Kinnard's survey of more than one hundred army general-grade officers in the Vietnam War, the majority revealed a "lack of enthusiasm" for Westmoreland's strategy.[11] Yet the respondents remained anonymous and their critiques came in retrospect. Even when army officers did openly voice their reservations about the MACV strategy, Westmoreland ignored them.

In 1966, the army published *A Program for the Pacification and Long-Term Development of South Vietnam,* more commonly known as PROVN. The PROVN report, led by Army Chief of Staff Gen. Harold K. Johnson, urged the U.S military to focus more on pacification, supported with evidence of the ineffectiveness of search and destroy missions. Westmoreland rejected the report. To quote Philip Davidson, "Westmoreland's rejection of PROVN was the old factor of 'not made here,' that is, not produced by himself and his staff in Saigon."[12]

The decades leading up to U.S. military intervention in Vietnam had brought numerous opportunities for interservice disputes. The U.S. Marine Corps was usually in the center of what often proved to be vociferous debates. Deeply imbedded in the culture of the Marine Corps is a sense of paranoia and vulnerability.[13] The history of the Marines in the twentieth century, most notably after World War II, is largely one of institutional survival, which helps to explain the defensive nature of Marine leaders in the Vietnam era.

The Marines who would become commanders in the Vietnam War had experienced in the previous decades an interservice debate that had almost completely depleted the Marine Corps. In the years immediately after World War II, the creation of the U.S. Air Force incited concerns in the U.S. Navy and U.S. Marine Corps over possible reductions in funding, strength, and size. In 1946, Carl Spaatz, the commanding general of the Army Air Forces (soon to be the U.S. Air Force), argued that a separate air force would prove capable of operating all air operations in future wars. Gen. Dwight D. Eisenhower, who had taken over as the army chief of staff, agreed with Spaatz, and went further in pushing for "severe restraints" on the size of the Marine Corps to allow for more financial allocations for strategic air power.[14] Although the Marines faced possible extinction in the late 1940s, the ensuing Korean War briefly reversed their fortunes. The successful amphibious assault at Inchon in September 1950 and the hallowed Chosin Reservoir campaign later that year gave the Marines some political capital in the years following the Korean War.

Moreover, in the late 1940s, carrier-based air and amphibious assaults seemed obsolete with the possibility of nuclear war on the horizon. With the increasing threat of nuclear war with the Soviet Union in the 1950s, the Eisenhower administration emphasized bolstering the newly established U.S. Air Force's strategic bombing capabilities while reducing the conventional elements of the army, navy, and Marine Corps. From 1958 to 1960, the Marine Corps saw its budget and overall strength decline. The 170,621 Marines who comprised the Corps in 1960 influenced then commandant Gen. Randolph McCall Pate to declare that a force that small could handle "only minor crises."[15] The Marine Corps pinned its hopes of institutional survival on the United States avoiding a major war with the Soviet Union.

President John F. Kennedy's "Flexible Response" strategy revived the fortunes of the Marine Corps. Historian Allan Millett notes that under this strategy, the Marine Corps benefited from the Department of Defense's budgeting system, which placed the Marines' finances into a "General Purposes Forces" category that allocated billions of dollars to all services.[16] Still, although the danger seemed past for the Marine Corps, it was fresh in the minds of Vietnam-era Marine leaders, who had experienced the interservice upheavals firsthand.

Gen. Wallace Greene Jr. represented the Marine Corps in interservice disputes in the months leading up to combat involvement in Vietnam. From 1 January 1964 until 31 December 1967, Gen. Greene served as the twenty-third commandant of the U.S. Marine Corps. Greene had served in the Pacific in World War II. During the 1950s, he worked in numerous capacities at Marine Corps schools in Quantico as well as becoming staff special assistant to the joint chiefs of staff (JCS) for National Security Council affairs. Greene rose through the ranks, landing at Camp Lejeune in 1957 as a brigadier general in command of the Marine Corps base in North Carolina. Greene became assistant chief of staff of the Marine Corps and ultimately, in 1960, chief of staff, a position he held until Kennedy nominated him for commandant in late 1963. Thus, Greene was no stranger to the behind-the-scenes politics that took place among top military leaders in Washington. In 1964, when the JCS pondered who should join Westmoreland's MACV staff, Greene disapproved when the top military positions in Saigon went to army officers.[17] The JCS also discussed who would succeed Adm. Harry Felt as commander in chief, Pacific (CINCPAC), a title traditionally held by a U.S. Navy officer. Whoever became CINCPAC would serve as the intermediary between Washington and MACV, thus making the position a prestigious and influential one. Greene angered the Department of the Navy when he voted for an air force officer to succeed Felt. Although the navy convinced McNamara to give the CINCPAC post to Navy admiral Ulysses S. Grant Sharp, animosity between Greene and Chief of Naval Operations Adm. David McDonald continued over the succeeding years.[18]

In 1964, during deliberations over the strategy the U.S. military would use when combat troops arrived in Vietnam, Greene pushed the JCS to support the enclave strategy and a sizeable Marine presence to occupy the enclaves. The joint chiefs disregarded Greene's proposal. Army Chief of Staff Gen. Harold Johnson cautioned his colleagues that an expanded combat role for U.S. Marines could lead to excessive casualties.[19] Within the first month of the landing of the first contingent of Marines at Da Nang in March 1965, Greene voiced his frustrations to Johnson about relegating IIIMAF forces to static defensive positions.[20] Until his time as commandant expired in 1967, Greene continued to show vocal support for the enclave strategy, much to the chagrin of many of his JCS colleagues and Westmoreland.[21]

By June 1965, Westmoreland had released the Marines from their defensive postures in I Corps. Despite MACV and JCS overtures to eliminate the enclave strategy, the Marine Corps stayed the course with Greene's plans. As the Marines began to expand their areas of operation during the summer, Walt took command of IIIMAF. In agreement with Greene, Walt began implementing the "inkblot," or enclave strategy, for the war in I Corps. Throughout 1965, the Marines gradually expanded the enclaves they had established at Da Nang, Phu Bai, and Chu Lai. As the population under Marine control increased with the expansion of the enclaves, IIIMAF provided civic action materials and performed pacification duties in the villages. Walt wanted to take advantage of the opportunity to send Americans to the villages to interact with civilians in hopes of gaining their support. Westmoreland, irritated with the Marine strategy, wrote in late December 1965, "The Marines have become so infatuated with securing real estate and in civic action that their forces have become dispersed and they have been hesitant to conduct offensive operations except along the coastline where amphibious maneuvers could be used with Naval gunfire support which is available."[22]

Westmoreland argued that the Marine fixation on the areas within their enclaves allowed the enemy to tighten its grip on the outlying areas, further undermining the Marines' ability to expand their tactical areas of responsibility. To ensure that all American infantry units achieved high enemy body counts, Westmoreland instituted "battalion days in the field," in which MACV ordered U.S. military battalions to execute a given amount of patrols away from their base camps. Any operation dedicated to searching for main force units counted as a battalion day in the field; MACV denied the Marines' request to count pacification operations in this category. In November 1965, Westmoreland dispatched MACV operations officer Brig. Gen. William DePuy to I Corps to assess the military situation. After his visit, DePuy suggested to Westmoreland and IIIMAF headquarters that Marines conduct at least two multibattalion operations against the enemy per month.

If Walt had openly defied Westmoreland, *Time* magazine's "Man of the Year" for 1966, it could have resulted in the MACV commander (Walt's boss) relieving him of command. Walt was always polite and proper when dealing directly with Westmoreland. He never publicly challenged his boss's orders during his IIIMAF command tenure. Walt's generally cooperative behavior earned him two nicknames among his Marine colleagues: Silent Lew and Big Dumb Lew. Walt never knew about the latter. Victor Krulak once screamed at Walt, "Goddamit, Lew, don't ask them. Tell them."[23] Although Walt believed Westmoreland's strategy had its faults, he obeyed the command structure that had placed MACV headquarters in charge of the Marines.

Walt initiated a multipronged strategy in I Corps. During his tenure as IIIMAF commander, from 1965 to 1967, the Marines focused on pacification in the rural areas, small-unit counterguerrilla operations in the countryside, and large-unit operations against enemy main force units. In August 1965, Operation Starlite featured the Third Battalion of the Third Marine Regiment launching a conventional attack with copious close air support, vertical assaults with helicopters, and amphibious beach landings to envelop the First VC regiment, a main force unit, south of Chu Lai. Just one month later, the same Marine regiment participated in another conventional assault during Operation Piranha. By the end of 1965, the Marines had added another large-unit operation to their short history in Vietnam, engaging the First VC regiment again during Operation Harvest Moon.[24] In 1966, after numerous large-unit Marine operations, in addition to the U.S. Army's first major engagement with the NVA in the Ia Drang Valley, the U.S. military spent much of the year building up its resources and manpower while protecting the newly created installations and supply routes that accompanied the buildup.

The year 1966 also saw a major political event in I Corps that diverted IIIMAF's attention from CAPs and "the other war." In March, the Nguyen Cao Ky government in Saigon removed Nguyen Chanh Thi from his command as the South Vietnamese I Corps commander. Thi had gained the deep respect and loyalty of his South Vietnamese subordinates in I Corps. The ensuing political crisis neutralized any gains the Marines had accomplished in I Corps, including the expansion of CAPs. In an attempt to preserve order in the cities, the Marines diverted their attention from their field assignments to the political crisis. Walt recalled that "most of what we had accomplished in almost a year appeared to collapse."[25] Loyalties within the South Vietnamese military were divided, with the Vietnamese Air Force aligning with the government, largely due to Ky's dual position as president and chief of staff of the air force. The Vietnamese Army, however, including the territorial forces, sided with Thi. Within one day after Thi's removal, protestors loyal to the former I Corps commander lined the streets of Da Nang to voice their opposition.[26] By 19 March, Buddhists in Da Nang had begun showing their disapproval of Thi's removal, which ignited a separate protest in the imperial city of Hue.

There, local university students began protesting, gradually shifting their pro-Thi remarks into anti-American pronouncements. In early April, the South Vietnamese government declared that Da Nang was in the hands of enemy sympathizers. Therefore, Saigon mobilized the local Vietnamese Marine Corps, which had aligned itself with the government. In response, the South Vietnamese army factions near Da Nang that had denounced Thi's removal prepared to dispatch units to meet the oncoming Vietnamese Marines. Ultimately, the two leaders of each branch of the South Vietnamese military met and ordered their respective forces to stand down, much to the relief of the thousands of Marines who were ready to squash both factions to secure Da Nang.

Just as the political crisis in Da Nang was subsiding, the NVA launched attacks near the DMZ. These attacks, which continued unabated for the next year, forced Walt to transfer his attention and IIIMAF resources to northern I Corps. July saw seven of the Marines' eighteen available infantry battalions dispatched to the DMZ. By 1967, MACV had placed thousands of army personnel in I Corps to occupy the regions in the southern corps provinces abandoned by the Marines, who had moved closer to the DMZ. With seventy thousand troops at his disposal at the end of 1966, Walt simply could not afford to focus on pacification while the NVA was wreaking havoc near the DMZ. Losing South Vietnamese territory to the NVA surely would have had a negative impact on the historical prestige of the Marine Corps. Although they had enough manpower and resources to repel the NVA attacks, the Marines could not fight the conventional war and still effectively protect all populated areas within their enclaves.[27] The Marines' new focus on the DMZ played into the hands of the NVA, whose policies during 1966 aimed to disperse American forces across I Corps to achieve gains in the guerrilla campaign.[28]

By March 1968, thirty thousand army soldiers had arrived in I Corps to accompany eighty thousand Marines, providing a ripe environment for interservice rivalry. Walt's successor, Lt. Gen. Robert E. Cushman, retained control of IIIMAF operations. However, the increased army presence in I Corps called for additional senior MACV officials to monitor the soldiers. In early 1968, Lt. Gen. William B. Rosson arrived in I Corps as the commander of MACV Forward at Phu Bai, just south of Hue. Much to the annoyance of the Marine Corps, the MACV Forward commander controlled the northern half of I Corps, which included the Third Marine

Division, the army's First Air Cavalry Division, and all other Marine and army units in the area. The southern half of I Corps was left to the First Marine Division. Although Rosson was subordinate to Cushman's overall IIIMAF command, Westmoreland had created a dual army–Marine Corps command structure in I Corps.

Army critics of the Marines' handling of the increasingly volatile situation in I Corps chastised IIIMAF for its failure to construct static defense bases along the DMZ. MACV leaders blamed the situation in northern I Corps on the Marines' ignorance of intelligence reports on NVA movements.[29] In 1968, IIIMAF perceived that the commander of the army's XXIV Corps in northern I Corps in 1968, Lt. Gen. Richard Stillwell, strongly opposed CAPs, so IIIMAF decided not to place any more CAPs in the army commander's area of operations, between Hue and the DMZ.[30]

Although IIIMAF did not necessarily enjoy the presence of the U.S. Army in its corps tactical zone, the extra soldiers allowed the program to grow through the tumultuous years of 1966 to 1969. In fact, it was during these years that the program peaked at 114 CAPs. Without army reinforcements to I Corps, the Marines would have been stretched so thin that perhaps the number of CAPs would not have grown at all. Moreover, the newly arriving army units in southern I Corps provided the accurate and timely artillery support that Marine batteries had provided before their move north to help the U.S. forces battling the NVA near the DMZ.

Various army generals spewed scathing critiques of the Marines and CAPs. Maj. Gen. Harry Kinnard, commander of the army's First Cavalry, argued that the Marines "just wouldn't play. They just would not play. They don't know how to fight on land, particularly against guerrillas."[31] Even the secretary of the navy, Paul Nitze, commented in 1966 that the Marines seemed to lack aggressiveness in Vietnam in that they failed to pursue large enemy units.[32] U.S. civilian officials in Vietnam also joined the army in criticisms of Marine strategy. From 1962 to 1969, Richard Holbrooke worked in several U.S. civilian positions in Vietnam, including an assignment with USAID focused on improving pacification and the position of assistant to the U.S. Embassy in Saigon. In the middle of his tenure in Vietnam, Holbrooke contended that the U.S. Marines practiced "shallow and naïve 'pacification' activities" in I Corps.[33]

U.S. Army major general DePuy, one of Westmoreland's most trusted aides at MACV, held true to the army's belief that success in Vietnam depended on overwhelming firepower and mobility. When Westmoreland appointed DePuy commander of the First Infantry Division in 1966, the former MACV aide brought the focus on main force units with him to his new assignment. After his stint with the First Division, DePuy became special assistant for counterinsurgency and special activities (1967–1969), a curious assignment since he saw little value in counterinsurgency. DePuy's biographer contends that the army general embodied the tongue-in-cheek caricature of the Vietnam War: "Don't just sit there. Go out and kill something!" When U.S. combat troops entered Vietnam, "from then on," DePuy assessed, "pacification was secondary."[34] Reviewing the Marines' style of war in I Corps, DePuy claimed that "the Marines came in and just sat down and didn't do anything. They were involved in counterinsurgency of the deliberate mild sort."[35] DePuy recognized that the U.S. military was generally ill prepared for counterinsurgency, but he continued to advocate the war of attrition via search and destroy missions.

A few months after the U.S. Army's arrival in I Corps, FMFPAC commander Victor Krulak, the most outspoken Marine opposing Westmoreland's strategy, lashed out at the army. As FMFPAC commander, Krulak oversaw the logistics and training for Marines in the Pacific, but he had no operational control over them. In one of his final speeches before retirement in the summer of 1968, Krulak charged the army with attempting to abolish the Marine Corps.[36]

After serving in World War II and the Korean War, Krulak ascended the Marine ranks, in the early 1960s serving as special assistant for counterinsurgency and special activities in the Kennedy administration. Krulak was appointed thanks to a résumé indicative of his innovative thinking in addition to his personal history with President Kennedy. In World War II, Krulak advocated the construction of landing craft with a retractable bow, helping to inspire the creation of landing craft, vehicle and personnel (LCVP), or Higgins boats, as America's official amphibious landing craft in the war. In 1948, Krulak, who foresaw the usefulness of helicopters in combat, inspired the Marine Corps to execute the first helicopter assault maneuver in history. Also assisting in Krulak's appointment as counterinsurgency assistant was the personal relationship Kennedy and Krulak had established during World War II. In 1943, as a navy lieutenant in command of a torpedo boat, John Kennedy rescued Krulak and his fellow Marines before their landing craft sank. To show his gratitude, Krulak promised Kennedy a bottle of whisky if they ever met again. In 1962, in need of a counterinsurgency assistant, Kennedy gave the post to Krulak—who sent the president a bottle of whisky.[37]

During the American advisory effort, the JCS dispatched representatives from all four military branches to Vietnam to assess the military situation. Arriving in Vietnam in early 1963, Krulak and his joint chiefs colleagues agreed with MACV officials that the battle of Ap Bac, now seen by many scholars as a disaster for the ARVN, was an enormous success.[38] Between 1962 and 1964, Krulak visited Vietnam eight times. After labeling Ap Bac a success, Krulak began to argue for a counterinsurgency strategy in Vietnam. At a speaking engagement at the Naval War College in 1962 and in a *Marine Corps Gazette* article in 1963, Krulak assured his audience that warfare in Vietnam would entail gaining "hearts and minds" instead of ground.[39]

Instrumental in Krulak's shift in strategic focus was his meeting with Sir Robert Thompson, the British counterinsurgency expert and chief architect of Britain's winning strategy over guerrillas in Malaya. Thompson convinced Krulak that protecting the people constituted the most important component of a successful counterinsurgency. Krulak first noticed a problem with MACV strategy in March 1965. Relegated to defending the airbase at Da Nang, Krulak acknowledged, "This was never going to work. We were not going to win any counterinsurgency battles sitting in foxholes around a runway, separated from the very people we wanted to protect."[40] By late 1965, Krulak was pessimistic about the outcome of conventional war in Vietnam. Instead, like most Marine commanders in Vietnam, Krulak pushed for the enclave strategy.

Throughout 1965, Krulak collected his thoughts on paper, hoping to present his findings to high-ranking military and civilian officials prominently involved in constructing strategy. Krulak's paper, "A Strategic Appraisal," argued that enemy main force units had already figured out how to combat effectively America's war of attrition. Based on the U.S. Army's battle with NVA main force regiments in the Ia Drang Valley in November 1965, Krulak concluded that the enemy had learned to fight American grunts in close quarters, rendering American air and artillery support useless. The war of attrition, according to Krulak, played into the hands of Hanoi's leaders. Conventional operations would result in substantial American casualties, and Hanoi knew that American body bags would "erode our national will and cause us to cease our support of the GVN."[41] The attrition ratio that Saigon and Washington advocated would not favor the United States in the long run, Krulak believed, because the Communists had a seemingly inexhaustible supply of combatants. "There is good basis," Krulak argued, "for concluding that a strategy built around manpower attrition promises us nothing but disappointment." After acknowledging that targeting the enemy's main force units did have some merit when the number of troops favored the United States, Krulak then pleaded that "if the enemy cannot get to the people, he cannot win, and it is therefore the people whom we must protect as a matter of first business." Krulak contended that the U.S. military had to embrace pacification and focus on the security of the South Vietnamese people. The FMFPAC commander also highlighted the political ramifications of the war. Except for the montagnards, with their individual tribal interests, Krulak maintained that the villagers had little if any loyalty beyond their hamlets. "It is doubtful if one villager in a thousand has ever heard of Premier Ky," Krulak argued, "or who even knows who his own Province Chief is." He concluded by expressing his conviction that implementation of his proposal would bring victory, but ignoring it would carve a pathway to defeat.[42]

Krulak's paper garnered support from Wallace Greene, but nobody else in high-ranking positions gave it much attention. In January 1966, Greene echoed Krulak's thoughts, postulating that the Communists would win the war of attrition even though "their casualty rate may be fifty times what ours is." Greene also took a jab at Westmoreland, contending, "This is a thing that apparently the Army doesn't understand."[43] According to Krulak, because he knew that Westmoreland would dispute the paper, the disgruntled Marine bypassed him, taking his thoughts directly to Washington. In the summer of 1966, Krulak scored a meeting with Lyndon Johnson in the White House. After less than one hour of discussion, an uninterested Johnson escorted Krulak out of his office. As Krulak remembered, Johnson "got to his feet, put his arm around my shoulder, and propelled me firmly toward the door."[44]

Johnson's rejection did not deter Krulak; he persisted in his efforts to be heard by other top Washington officials. In 1966 and 1967, Krulak presented his thoughts on the war to Secretary of Defense Robert McNamara. In a conversation they had in the spring of 1966, the defense secretary agreed with the FMFPAC commander on most of the issues they discussed. Yet Krulak believed McNamara had not pledged his wholehearted support. Between May 1966 and January 1967, Krulak wrote two letters to McNamara reiterating the necessity of removing the guerrillas from the villages. In a loose reference to CAPs, Krulak contended that U.S. forces must focus on developing village militias and guide the villagers toward a more stabilized life in their hamlets.[45] McNamara never responded to Krulak's letters. In his memoir *In Retrospect*, McNamara blames himself for failing to initiate a proper debate between the army and Marines about their differences.[46]

Shortly after arriving in South Vietnam, Walt admitted he did not have an immediate answer for how to win such a complex war. The IIIMAF commander conceded that the Marines had much to learn about the war when they first arrived. Yet as the Marines extended their areas of operation, they interacted with the South Vietnamese people and gained a better understanding of their needs. After a few months in I Corps, Walt noted: "It was a new kind of war we were in, where concern for the people was as essential to the battle as guns or ammunition, where restraint was as necessary as food or water. It was a war requiring a stronger discipline than it took to seize Mount Suribachi, and a war filled with new problems, demanding new solutions." According to Walt, CAPs represented the most successful innovation of the Marines in Vietnam. Walt ultimately found that civic action and CAPs offered the best formula for victory. "The struggle was in the rice paddies," Walt asserted, "in and among the people, not passing through, but living among them, night and day, sharing their victories and defeats, suffering with them if need be, and joining with them in steps toward a better life long overdue."[47]

After Walt's tenure as IIIMAF commander, he addressed the strategic situation in Vietnam at several speaking engagements. Lacking Westmoreland's command, Walt unleashed negative comments about the MACV commander's choice of strategy, although he never mentioned him specifically. "To me," Walt said, "it makes no sense to expend all our effort thrashing around the boondocks looking for the regulars while the VC are tightening their hold on the populated areas."[48] Walt maintained that search and destroy would not bring victory by itself. Rather, the focus must remain on the hearts and minds of the Vietnamese people. As Walt perceived the war, two missions existed for the U.S. military: "the destructive mission and the constructive mission."[49] To bring victory, both had to be achieved simultaneously.

In 1966, Westmoreland wrote in his yearly MACV report that CAPs "deserve the greatest credit and admiration."[50] Still, he saw many faults with CAPs. As he explains in a book written after the war, one of the primary reasons Westmoreland offered some strategic leeway to the Marines was to avoid an "interservice imbroglio." In formulating strategy, Westmoreland feared that frequent interaction between the Americans and South Vietnamese would invoke xenophobic feelings, likely resulting in "unfortunate incidents." Yet Westmoreland contends that if he had been given unlimited manpower during the war, he could have placed Americans in most South Vietnamese districts, allowing the troops to familiarize themselves more intimately with the people. According to Westmoreland, this would have given the United States an advantage, allowing the troops to more effectively identify insurgents and protect the civilian population. Thus, Westmoreland contradicts himself, on one hand arguing against implanting troops near the population, and on the other proposing that placing American military personnel near the population would have extracted vital intelligence information. Westmoreland acknowledges that the program achieved "noteworthy results" in counterinsurgency, but argues that he "simply had not enough numbers to put a squad of Americans in every village and hamlet"; implementing a CAP-like strategy across South Vietnam "would have been fragmenting resources and exposing them to defeat in detail."[51]

Placing troops in every village would have enabled the Americans to interact with the people, but as Westmoreland asserts, for that he would have needed millions more men.[52] Westmoreland's primary concern was the larger enemy main force units. When he commenced search and destroy in 1966, Hanoi answered with a similar aggressive strategy. While insurgent military commanders continued to adhere to guerrilla warfare, choosing battle on their own terms, the aggressiveness of North Vietnamese forces resulted in mounting casualties for the DRV. Combined U.S. and ARVN search and destroy operations succeeded in disrupting enemy operations, driving North Vietnamese forces from base camps and denying them provisions.[53] The NVA had matched the larger American units with battalion and regiment-sized operations of their own. By 1967, Hanoi had shifted toward a more elusive strategy, until the Tet Offensive one year later. Westmoreland made overtures to block infiltration routes by sending three divisions into Laos, but after realizing he could not spare the troops—and also knowing Johnson's likely rejection of such a strategy—he abandoned this idea. According to the MACV commander, "Nobody ever advanced a viable alternative that conformed to the American policy of confining [the] war within South Vietnam. It was, after all, the enemy's big units—not the guerrillas—that eventually did the South Vietnamese in."[54] The Marine Corps would beg to differ. William Corson, who served as the program's first director from 1967 to 1968, believed Westmoreland and the U.S. Army failed to support the program because if they threw their full support behind a successful Marine Corps counterinsurgency effort, the army's shortcomings would become apparent.[55]

In 1968, just one year after leaving his last post in Vietnam as the director of the program, William Corson published a scathing critique of American strategy in Vietnam. Corson blasted search and destroy operations for the disregard participating American units had for the Vietnamese people. Corson was much more direct and politically incorrect in his criticism of the U.S. Army than Walt. Corson detested the "golden ghettoes," or base camps, the army had constructed. The army soldiers' responsibility, according to Corson, ended at the gates of the base camps. He asserted that Marine base camps were placed either adjacent to or within a hamlet area. "In essence," Corson argued, "many marine battalion base areas became another hamlet in the villages in which they were located."[56] Unsurprisingly, Corson gave glowing appraisals of CAPs, and specifically of the Marines stationed in the villages. Corson served as the director of the program for less than a year, and the reasons for his abrupt departure remain unclear. From the abrasive and aggressive attack put forth in his book, one can gather that Westmoreland's military strategy played a significant role in his decision to leave the program and Vietnam altogether.

Throughout the war, top IIIMAF commanders sought ways to increase the presence of CAPs in I Corps. In 1967, Corson argued that obtaining ten thousand men for the program would provide enough men for the seven hundred villages of I Corps, thus bringing the war to a manageable level within one year.[57] With only one thousand Marines providing security in April 1967, the Marines would need a massive overhaul of the program, but the NVA attacks near the DMZ prevented that from happening. In 1967, the general security of I Corps, most notably in the northern provinces, was not showing signs of improving.

As had Westmoreland, Robert Komer, the initial civilian leader of CORDS, reversed his thoughts on the program throughout his short tenure. Although Komer occasionally commented positively on the program when visiting with Krulak, evidence shows that he also disapproved of CAPs. After Komer canceled a briefing on the program in late 1967, a MACV official spoke for the absent CORDS leader in saying "the Combined Action Program is too expensive to continue."[58] Komer also argued, like Westmoreland, that the U.S. military simply did not have enough manpower to greatly enhance the program. In January 1968, Komer suggested to then Marine commandant Gen. Leonard F. Chapman that IIIMAF reduce the number of Marines in CAPs to eight-man teams with more mobility, similar to the army's MATs. Yet in 1970, years after leaving his post as the head of CORDS, Komer wrote publicly that CAPs "were an imaginative and effective approach to hamlet security, though never built up very far."[59] IIIMAF commander Gen. Cushman resisted the idea, but he also acquiesced to MACV and CORDS requests and created MTTs. The mobile teams consisted of Marine infantry squads that trained non-CAP PF for a few weeks before moving on to the next territorial force unit. In 1968, program director Lt. Col. Byron F. Brady acknowledged a "very poor" relationship with CORDS and MACV.[60]

In 1970, CORDS representatives in I Corps voiced their concerns about CAPs and the general structure of the program. In March, Francis McNamara, the political advisor to the commanding general of XXIV Corps, the army command that had moved to I Corps, directed a survey of the 114 CAPs. McNamara reported to XXIV commander Lt. Gen. Zais that while the villagers involved with CAPs appreciated the presence of the Marines, the program lacked coordination with CORDS. In his letter, McNamara suggests that the program should adhere to the CORDS single-manager concept of pacification.[61] The program had a command structure separate from the CORDS chain of command. Thus, from the CAGs down to the CAPs, the Marines coordinated with but were not subordinate to CORDS advisors. As McNamara highlights, the commanding officers of the combined action companies (at the district level) and combined action groups (at the province level) often held the same rank as the CORDS advisors at the district and province levels. The commander of the program, Col. T. E. Metzger, refused to accede to McNamara's proposal of placing CAPs under the direction of CORDS. Metzger argued CORDS did not have the capability to direct or support small tactical military units such as CAPs.[62] Two days after Metzger sent the letter to Zais, XXIV Corps assumed operational control of the program, which by this date had changed to the Combined Action Force, upon which it was given separate command status. With a separate command status, the commander of the Combined Action Force became an operational commander with direct access to the IIIMAF commanding general, allowing him access to IIIMAF resources.

The summer of 1968 ushered in a change in command at MACV headquarters. Gen. Creighton Abrams, who had served as Westmoreland's deputy MACV commander, now controlled the reins of U.S. strategy in Vietnam. In what Lewis Sorley has called the "one war" approach, Abrams recognized the equal needs of combat operations, pacification, and advising elements to achieve a single objective: security in South Vietnamese villages.[63] Before Abrams's ascension to MACV commander, Westmoreland dispatched his deputy in early 1968 to lead the new MACV-Forward position in I Corps. As head of MACV-Forward in Phu Bai, Abrams gained control of Marine and army forces in the northern part of I Corps. In mid-1966, the NVA began assaulting I Corps just south of the DMZ. IIIMAF commanders diverted attention to the area near the DMZ. When Abrams arrived at MACV-Forward in 1968, the army had already increased its presence throughout I Corps to reinforce the Marines, spread thin from their focus on the DMZ. One of the reasons for the creation of MACV-Forward was the army's disapproval of Marine strategy in I Corps. In early 1968, the Marine base at Khe Sanh, under siege from the NVA, concerned Abrams. In a meeting with Marine division headquarters, Abrams showed signs of frustration with the IIIMAF commander's inability to answer direct questions about the tactical situation at Khe Sanh. As the Tet Offensive raged in Hue, Abrams also was dissatisfied with Cushman's apparent mishandling of artillery and air support for the Citadel in the heart of the imperial city.

With the assistance and advice of Ambassador Ellsworth Bunker and Komer's successor as CORDS director William Colby, Abrams set out to bolster South Vietnamese forces before the looming American withdrawal. The increased emphasis on pacification by bolstering the territorial forces, a necessary focus due to Vietnamization, resulted in fewer battalion-sized airmobile operations. As Westmoreland's deputy commander from 1967 to 1968, Abrams traveled throughout South Vietnam, gaining an awareness of the need to improve the territorial forces. By the time Abrams became MACV commander, population security, according to Sorley, had superseded body count as the primary measure of success.[64] Although CAPs seemed to adhere to Abrams's "one war" approach, Sorley never highlights the program's relationship with the new MACV commander and his strategy.

In early 1968, Gen. Wheeler approached Abrams about the possibility of assigning Krulak as deputy MACV commander once Westmoreland had departed. Krulak had thought about retiring, which he did later that summer, but possessing a spot at MACV headquarters would definitely force him to reevaluate his future plans. Without specifically mentioning Krulak, Abrams responded to Wheeler that "no Marine has the full professional military qualifications to satisfactorily discharge the military responsibilities of the office." In justifying his response about the Marines, Abrams noted to Wheeler, "I consider them less professionally qualified in the techniques and tactics of fighting than the U.S. Army, the Korean Army and the Australians. The Marines have in the main been slow to adapt innovations, tactics, techniques and devices which would make their forces more effective against a frequently cunning and clever enemy. They have not been imaginative in developing ways to optimize their strong points against the enemy's weak points."[65]

These derogatory comments from Abrams, who supposedly placed high priority on securing the population and bolstering the territorial forces, come as a surprise considering the program's dedication to executing the exact duties Abrams applauded. Until the creation of MATs and MTTs in 1968, CAPs represented the only American force dedicated to training the PF. A self-noted frequent traveler of all corps regions of South Vietnam, Abrams was unquestionably aware of the program's efforts and gains in I Corps.[66] By the end of April 1968, about one month before Abrams assumed command of MACV, 718 PF platoons with a total strength of more than 21,385 men existed in I Corps. The 85 CAPs in I Corps that same month featured an additional 2,153 PF members. Despite constituting only one-tenth of the overall PF strength in I Corps,

PF associated with CAPs accounted for one-third of the enemies killed or captured by the entire PF force in I Corps.[67] In 1969, when the program peaked at 114 CAPs, PF in the program accounted for 1,952 enemy KIAs, compared to the 3,411 killed by non-CAP PF, who vastly outnumbered the former.[68] Indeed, Abrams did not measure success by body count. However, the evidence shows that the training process for PF under Marine supervision was moving forward much more rapidly compared to those unassociated with the program.

Although many army commanders showed an obvious disdain for the Marines' counterinsurgency efforts, the army's American Division established a unit in I Corps that mirrored CAPs. In 1969, in conjunction with the accelerated pacification campaign, First CAG personnel assisted one of American's infantry companies in its combined operations with the PF. In this case, the combined army-PF unit created a defensive shield against the NVA, while the CAPs they protected concentrated on the VC in the area. The American's combined company was considered a part of the Infantry Company Intensified Pacification Program, which emerged in the midst of the accelerated pacification campaign. By December 1969, the American had two ICIPP, ultimately changing the name to CUPP units within the First CAG enclave.

One should not assume that army units in I Corps refused to work with CAPs. When the army began occupying the southern parts of I Corps, it provided the artillery and logistical support for CAPs in their area of operations. Marines involved in those CAPs applauded the army support they received during their time in the village.[69] Although many of the general-grade officers in the army disliked and in extreme cases criticized the implementation of CAPs, the army and Marine Corps units in the field worked together in an effective manner to help the program achieve its military goals, and in the case of the American unit, its pacification goals.[70]

From the commanders at MACV headquarters in Saigon to the highest-ranking military and civilian leaders in Washington, the impassioned proponents of the program never garnered full support for CAPs from anyone outside the Marine Corps. Neither of the American land-based services' commanders would acquiesce to criticism and willingly admit that their way of fighting the war was ill conceived and poorly executed. The program's foremost champions, such as Victor Krulak, believed the army attempted to liquidate CAPs. After years of interservice disputes and continual talk of depleting the Marine Corps before the Vietnam War, the success of the program offered its parent institution the chance to justify its relevancy. Krulak especially sought to prove that the Marines employed a viable and more productive alternative to the army's attrition-based strategy. Not all of the program's proponents were as vociferous or aggressive as Krulak. "Silent Lew" was tactful with Westmoreland during the former's tenure as IIIMAF commander. Moreover, Walt saw value in the conventional large-unit operations that MACV advocated. It is too simplistic to point to the army's strategic decisions or its commanders' perceptions of the Marine Corps and CAPs as the sole reasons for the program's disappointing growth rate. The war itself proved to be the chief instigator of the program's demise.

Conclusion

The driving core of Marine culture, even more than a sense of the past, is
its sense of future vulnerability. Every Marine is taught that the very existence of the Marines is always in danger.

—Thomas E. Ricks

The U.S. Marine Corps Command and Staff College in Quantico, Virginia, has produced numerous studies of the program, most of which offer
fervid appraisals of CAPs in Vietnam. In 2002, Maj. Curtis L. Williamson's study argued that the dispersal of CAPs throughout South Vietnam
would likely have preserved the country's sovereignty. His revisionist approach estimates that placing a CAP in every village of South Vietnam
would have required a "reasonable sum" of thirty-two thousand Americans and seventy thousand PF.[1] Overall, scholars examining the "what ifs"
of military history can make countless estimates and predictions about the
Vietnam War. In the end, however, one can only speculate how the Vietnam War would have turned out if the program had expanded to every
corps tactical zone in South Vietnam. Nobody knows exactly what would
have happened to the U.S. military effort in Vietnam if American war
planners had employed the CAP strategy on a much larger scale. However, historians and scholars can certainly research and write accurately
about what did actually happen. In the case of CAPs, many Americans
in the program left the war with a newfound respect for the Vietnamese
people.

In 1967, Corson proposed to Defense Secretary Robert McNamara plans to ultimately place four hundred CAPs in I Corps. Although the number of CAPs obviously never reached anywhere near that mark, the former program director was a true believer in the effectiveness of CAPs. To achieve that mark would have required more than forty-four hundred Marines and corpsmen in the villages alone. This high number of CAPs also would have necessitated the creation of new combined action companies, each manned with several Marines and PF. Moreover, the increase in CAPs would either have created new combined group headquarters or would have brought more Marines to the already existing ones. Not only would it have taken more manpower to accomplish this feat, but in a strategic sense it would have handcuffed IIIMAF's ability to fend off enemy main force units with Marine infantry and artillery that otherwise would have supported the CAPs.

The vast majority of program veterans agree with Corson's assessment of the effectiveness of CAPs. One of the exceptions is Edward Palm, who from July 1967 to January 1968 served as a Marine corporal in CAP "Tiger Papa Three" in Quang Tri province. After his return to the United States, Palm declared in an article in the *Marine Corps Gazette* that his CAP "had been a failure that can be attributed almost totally to intercultural misunderstanding."[2] In Al Hemingway's collection of oral histories of CAP veterans, Palm expands on his earlier article, arguing that there were not enough Marines in Vietnam with the "intelligence and sensitivity" to make the program successful on a larger scale.[3] Palm's assessment further supports the truism that every American in Vietnam, and more specifically every member of the Combined Action Program, had his own experiences and perceptions of his individual involvement in the war.

As this book has addressed, many of the leading Marine commanders during the war despised the U.S. Army, Westmoreland, and MACV in general for their strategy. Throughout history, the U.S. Marines have always fought with a chip on their shoulders. In addition to constantly attempting to live up to their "devil dog" and "leatherneck" reputation, which they embrace, they have always fought for institutional survival. The Vietnam War was no different. With the U.S. Army representing the only other land-based branch of the American military, the Marines felt the need to justify their individuality as a service. When the U.S. Army challenged the strategy they had built, Marines such as Wallace Greene and Victor Krulak, both of whom had experienced the downsizing of the

Corps in the decades before Vietnam, fought back with vigor. In light of the history of the U.S. Marine Corps in the twentieth century, it is easy to understand why Greene and Krulak were so defensive about the use of Marines in Vietnam. On the other hand, the U.S. Army had no institutional plans to extinguish the Marine strategy in Vietnam. After all, two U.S. Army generals held operational control over the Marines during the war, and both had the power to terminate CAPs. There was no attempt to do so, although there is equally no doubt that many of the top U.S. Army commanders loathed what the Marines were doing in I Corps. Westmoreland even relegated the primary unconventional element of the U.S. Army, the Green Berets, to border surveillance rather than the counterinsurgency and counterguerrilla tactics they had employed for years. He never did that with CAPs. The only changes to the employment of CAPs came from the program itself when it changed to the mobile concept. Even Walt, credited as perhaps the biggest influence in implementing the Marines' way of war in Vietnam, noted the importance of annihilating enemy main force units. According to Walt, the Marines chose to participate in large-unit search and destroy missions because the IIIMAF commander believed that strategy had some merit in the war. More than any other factors, the manpower shortage and Vietnamization prevented the program from enlarging beyond its peak of 114 CAPs.

In 2003, thirty years after U.S. forces withdrew from Vietnam, U.S. and coalition forces, in one of the most dominant campaigns in military history, destroyed the conventional military forces of Iraq and deposed Saddam Hussein in just three weeks. However, the George W. Bush administration failed to plan for the insurgency that rapidly developed after the successful conventional assault. In the typical American military fashion of reacting to situations in an ad hoc manner, U.S. commanders had to formulate counterinsurgency doctrine that the army and Marine Corps had neglected over the previous thirty years.

In 2006, the army and Marine Corps collaborated to release a field manual on counterinsurgency, *FM 3-24,* a collection of essays written by about twenty primary authors, headed by retired army lieutenant colonel Conrad Crane. Six hundred thousand editors reviewed the manual before its publication. Many soldiers and Marines in Iraq with counterinsurgency experience provided suggestions and input before the final product hit bookshelves and websites. Recognizing the complexities involved in any counterinsurgency, Crane notes that the manual's purpose is to provide guidelines for counterinsurgents; it is not meant as a book of "cookie-cutter solutions, but it does contain a lot of good ideas, historical lessons and insights."[4] Within the first month of its posting on army and Marine websites, the field manual was downloaded more than 1.5 million times.

FM 3-24 discusses in detail the criteria for an effective counterinsurgency, shedding light on the importance of intelligence, developing and training indigenous forces, and leadership. Chapter 5, "Executing Counterinsurgency Operations," devotes several pages to combined action, including a tribute to CAPs in Vietnam as "a model for countering insurgencies."[5] The pages dedicated to combined action provide suggestions for units living with Iraqi civilians while training the indigenous forces. The guidelines given for combined action describe attributes similar to those of CAPs—unsurprisingly, considering the acknowledgment of the Vietnam program's relevance in the Iraq context.[6] The manual concludes that the U.S. element of a combined action unit must develop a positive relationship with local security forces and civilians. American military personnel, according to the manual, should learn from the indigenous forces about local customs, surrounding terrain, and possible insurgent hideouts.[7]

CAPs have become an increasing point of interest for books about the U.S. military during the Iraq War. One year before the cessation of U.S. combat operations in Iraq, Bing West published *The Strongest Tribe*. West, a former Marine whose book *The Village* recounts his time with a CAP in Vietnam, writes about the counterinsurgency accomplishments in Iraq that he believes have been overlooked. Frequently referring to his tenure with a CAP in Vietnam, West reveals how the Marines employed combined action in Iraq. CAP Marines in Iraq performed functions similar to those of their predecessors in Vietnam. The Americans shared command centers with the local forces and took them on daily patrols; they also offered civic action to the civilians in the towns they occupied.[8]

In 2004, the First Marine Division commander in Iraq, Gen. James Mattis, required that every battalion under his command contain a CAP. Mattis had gained a knowledge of and respect for CAPs in Vietnam through his relationship with Bing West. Despite Mattis's orders, however, only a few CAPs operated in Iraq. Moreover, Marine CAPs in Iraq did not receive anywhere near the level of support or guidance from the upper echelons of the Marine Corps as did the combined units in Vietnam.[9] The Second Battalion, Seventh Marine regiment established a CAP in Hit, the third largest city of Al Anbar province. The CAP, which carried the call sign "Golf-3," began training the local Iraqi security forces in late May 2004. The Golf-3 Marines had difficulty learning the local customs and language—the unit's predeployment training consisted of only two days of instruction on the mission. Yet, like the CAP Marines in Vietnam, the members of Golf-3 quickly learned basic military tactical terms to communicate with the indigenous forces. The CAP of the Seventh Marines completed its mission after only four months, leaving the Iraqi forces to guard the area on their own.[10]

Shortly after Golf-3 departed Hit, another CAP emerged from the Third Battalion, First Marine regiment. CAP India, remembers one of the Marines in the unit, "gained incredible insights into the hearts and minds of the Iraqi people." CAPs in Iraq, as in Vietnam, went on daily patrols with their indigenous counterparts in search of booby traps to disarm. Unlike in Vietnam, the primary booby trap in the open desert roads were improvised explosive devices (IEDs). Insurgents hid IEDs at strategic locations along U.S. logistical lines. On IED sweeps in CAP India, every Marine was assigned an Iraqi counterpart. The indigenous forces, with little training, were initially reluctant to serve alongside the Marines. To motivate them, the Marines held classes on patrol techniques and tactics, often followed by screenings of movies such as *Blackhawk Down* and *We Were Soldiers,* in addition to the HBO miniseries *Band of Brothers.*[11] By June 2009, the only two CAPs in Iraq had disbanded, and the implementation of Marine combined action died with the departure of Gen. Mattis and the First Marine Division.[12]

With the combat phase of the Iraq War over, the United States' attention now focuses squarely on the remaining troops fighting the Taliban in Afghanistan. The U.S. Marine Corps has employed the combined action concept in Afghanistan as well. Bing West's *The Village* has not only graced the Marine commandant's annual reading list, it has made its way into Afghan villages guarded by CAPs. The CAP Marines read "dog-eared copies" of the book and have named several of their outposts after CAP Marines killed in the Vietnamese village of Binh Nghia, the setting of *The Village.*[13] Mark Moyar has predicted that the use of combined action units in Afghanistan will likely never reach the scope of CAPs in

Vietnam. The reasons for this stem from the "force-protection" measures that typify U.S. military operations in Afghanistan. Moyar asserts that "Americans are required to operate in larger numbers, in closer proximity to each other, with multiple armored vehicles. We do not even attempt to control a large fraction of the village because they are inaccessible by armored vehicle or are too far from reinforcements."[14]

Only time will tell if the CAP concept in Afghanistan was successful. Although CAPs in Vietnam may provide a solid framework upon which the U.S. Marines today can build, the wars in Southeast Asia and Afghanistan are separated by decades of technological and cultural change in America's military. Today's military is armed with global positioning devices, more efficient and advanced medical treatment and facilities and, as Moyar's words reveal, an increasing reliance on nearby armored vehicles and plenty of reinforcements via air and ground. Moreover, the geography and culture of Afghanistan present new challenges for today's military in adapting counterinsurgency techniques. Yet despite the lapse of time between the two wars and their different circumstances, one overarching factor remains constant: the critical need of the American contingent to respect the indigenous civilians and adapt to their culture and way of life.

The program experienced some success at particular points in the war; as it grew, the increased security of the hamlets in which CAPs operated definitely succeeded in eradicating a large portion of the VC presence in the villages. However, as the program peaked in units and manpower, the overall American military establishment in Vietnam began to dwindle, and so did the CAPs.

During the summer of 2007, I studied for three weeks in Southeast Asia, courtesy of the Vietnam Center at Texas Tech University. I traveled to both urban and rural areas of Vietnam, visiting sites such as Ap Bac and Khe Sanh, whose very names stir painful memories for Americans who experienced the Vietnam War firsthand. I met with former NVA soldiers, current Communist Party members of the government, and university students. Coming from a country where the word *Vietnam* conjures up the most controversial war in U.S. history, I was shocked to find that the Vietnamese today have largely forgotten about their "American War." Vietnamese college students, I discovered, were very knowledgeable about and interested in early revolutionary U.S. history, just as Ho Chi Minh had been in the decades leading up to the August Revolution of 1945. Yet the students were mum when it came to the Vietnam War.

Of course, Vietnamese Americans who fought for South Vietnam remember the war vividly. Many of them despise the Communist government that exists today and would not welcome the thought of their children conversing with the son or daughter of a former North Vietnamese participant in the war.

In some cases, American scholars and veterans are still fighting the Vietnam War by attempting to revise history, searching for the winning strategy that never was. Much of this revisionist history places the U.S. military under a strategic microscope yet forgets the enormous South Vietnamese military. Many scholars have closely examined U.S. strategy in Vietnam, attempting to identify what went wrong, and in doing so to suggest what American forces could have done to change the outcome of the war. However, many of these works exclude any detailed evaluation of the South Vietnamese military. Some scholars are actually making the same mistakes the U.S. military in general made decades ago—trying to win the war for South Vietnam.

As was true of the entire U.S. military effort in Vietnam, CAP Marines tried to win the war for the indigenous population. CAP Marines commandeered the vast majority of military activity in the villages, believing that was the only way to keep the village from falling into VC hands. The Marines could not rely on the PF to execute and complete their military responsibilities. Nor could they fully trust all of the thousands of villagers in a CAP village. The Marines knew that at a moment's notice, a sizeable VC force, buttressed by some of the village civilians, could attack their CAP unit of ten Americans and a few dozen village militiamen. Although respect between many of the Americans and villagers gradually increased over time, the Marines and corpsmen ultimately could trust only each other.

Unlike the general U.S. military effort, CAP Marines tried to win the war in the villages. The program was the war's largest American counterinsurgency program that sent military personnel to live in villages for the duration of the U.S. combat phase of the war. The program had an unmatched dedication to training the PF and providing security for civilians in their home villages. CAP Marines did not roam from one village to another to train different militia units. Nor did they forcefully relocate civilians to distant and foreign areas. Each CAP stayed in one village, trained the same PF platoon, and interacted with the same civilian population. Living in the villages brought numerous personal, cultural, and military challenges for the Americans, some of which proved difficult to overcome. Yet in facing these challenges, many of the Americans transformed their perceptions of the Vietnamese people. In CAP villages, Americans witnessed firsthand that the Vietnamese were neither emotionless nor uncaring toward the horrors of war. Like the Americans, they too experienced intense fear, sorrow, and pain. As this book also has shown, the Marines and corpsmen understood that the villagers simply did whatever was necessary to protect their families. In short, serving in a CAP humanized the villagers for the Americans. In the eyes of the Marines and corpsmen, they were no longer "gooks." They had become Vietnamese.

Acknowledgments

I owe a sincere debt of gratitude to numerous people in the history department at Texas Tech University for helping me complete my doctorate degree and this book: Laura Calkins, Lynne Fallwell, Randy McBee, Justin Hart, and Patricia Pelley. Each of them offered either personal or academic support far beyond the realm of their job descriptions. I would also like to thank Jim Reckner for his encouragement over the years to continue my pursuit of military history. I will forever treasure our private conversations about the subject and profession of history, a truly enlightening experience. I am also grateful for Larry Berman's advice and suggestions on preparing this manuscript for publication.

Ron Milam deserves the utmost credit for propelling me through graduate school and my manuscript. He always put my concerns ahead of his own. Moreover, Ron's office was my main stop for nonhistory conversations, providing much-needed social relief. Ron is perhaps the most popular professor among undergraduate and graduate students alike at Texas Tech. His immense knowledge of the Vietnam War is surpassed only by his respect and appreciation for his students.

Several employees at the Vietnam Center and Archive at Texas Tech also deserve my recognition. During the summer of 2007, I studied abroad in Vietnam, courtesy of the Vietnam Center. This life-changing experience would not have happened without the Vietnam Center. Le Kang Khanh was an absolute lifesaver during our travels, spearheading linguistic transactions with the Vietnamese. Steve Maxner, the director of the Vietnam Center, organized the trip and for the past five years has provided assistance with research and letters of recommendation. Kelly Crager's advice on interviewing CAP veterans was an enormous help. Also, I would like to thank the employees of the Vietnam Archive, most notably Amy Hooker, for their research assistance.

I would also like to thank Peter Worthing at Texas Christian University, where I completed my undergraduate and master's degrees. I credit Peter with helping me make my choice to turn a love of history into a career. He graciously took much time out of his schedule to improve my skills as a historian in training. His unmatched love and passion for history is infectious, which has an enormous impact on students who have had the fantastic opportunity to work with him.

In 2010 and 2011, I had the distinct honor of interviewing veterans of the Combined Action Program at their annual reunions. I was blown away by their sincerity, friendliness, and generosity. A special thank you to Fred Caleffie, Michael Noa, Jose Molina, Jay Baxter, Mike Murphy, Paul Kaupus, Gary Evins, Pat Morris, Jerry Gonnell, Robert Averrill, Richard Thunhorst, Bryan Haddock, Bob Nation, Ron Schaedel, Bill Bennington, Mike Smith, Tom Morton, and Bill LeFevre and his wife, Sandy, for making my experiences at the reunions memorable. I also am greatly indebted to Nick Duguid, Robert Holm, Tim Duffie, Rick Groulx, Tom Harvey, Douglass Reed, Roger Marty, Channing Prothro, Robert Hall, and Ron Titus for providing photographs.

A special thank you also to the U.S. Marine Corps Heritage Foundation, which awarded me with a fellowship that allowed me to travel to Washington, DC, and Quantico, Virginia, to complete my research. The staff at the Alfred Gray Research Center offered timesaving advice and suggestions, maximizing the efficiency of my research visits.

It has been a pleasure working with the University Press of Kentucky. Everyone at the press involved with publishing this book exuded much-appreciated alacrity. Press director Steve Wrinn expressed enthusiasm and confidence in my work, and Allison Webster, assistant to the director, exemplified promptness and professionalism throughout the publication process. Kyle Longley and Kathryn Barbier read the manuscript and provided perceptive commentary that improved the final manuscript.

Last but certainly not least, I would like to thank my family. My parents, Dan and Kari Southard, have wholeheartedly supported my choice to become a historian. After three years as a business major, I chose to prolong my undergraduate education in order to receive enough history credits to apply for graduate school. My parents never even remotely questioned that choice. Many of my own students' desires to attend graduate school in history are undermined by parents who want them to be doctors or lawyers. My parents have given moral and financial support to allow me to achieve my goal of becoming a historian. Words cannot describe how much they mean to me and to this project. Finally, my wife, Rachel, sacrificed a great job and being closer to her friends and family in Arlington, Texas, to come to Lubbock so that I could teach at Texas Tech and research more efficiently at the Vietnam Archive. The long hours I spent writing and editing this book depleted our time together. Although frustrated at times, she never wavered in her support. It does not matter how many times I tell her: she can never know how grateful I am to her.

APPENDIX

Historiographical Essay

The historiography of America's involvement in the Vietnam War spans several decades and features a myriad of publications. Scholars have exhaustively dissected the social, political, economic, and military aspects of the conflict. Despite these countless publications, CAPs remain largely overlooked.

Many of the more popular overviews of the war completely omit any explanation of CAPs.[1] Both George Herring's *America's Longest War* and Stanley Karnow's *Vietnam,* perhaps the two most popular American overviews of the war, do not mention CAPs in any capacity.[2] Even Lewis Sorley's *A Better War,* which emphasizes Gen. Creighton Abrams's supposed focus on "the other war," ignores CAPs.[3] Some of the surveys that do mention CAPs fail to provide any extensive description of the program.[4] Although several books dealing with the military or pacification elements of the war allot multiple paragraphs or pages to CAPs, a thorough academic analysis of the Marine Corps counterinsurgency program has not heretofore emerged.[5]

The earliest significant acknowledgment of CAPs in the historiography came from revisionist scholars seeking to debunk arguments from the orthodox school that strongly criticized America's involvement in Vietnam. As a *New York Times* reporter stationed in South Vietnam during the early 1960s, David Halberstam provided the groundwork for the orthodox school.[6] Halberstam stressed that U.S. policymakers, knowing little about Vietnam and its culture, rushed into supporting the unpopular government of Ngo Dinh Diem. By the time U.S. civilian leaders realized the severity of the situation in South Vietnam, their chance to abandon the region had vanished, thus forcing America into a quagmire. Chester Cooper's *The Lost Crusade,* published during the war, is a notable addition to the orthodox school. Cooper, one of the American representatives at the Geneva Conference in 1954, argued that the drive to spread American ideals to the rest of the world led the United States into a war it could not win.[7] More than forty years after Halberstam's publication, John Prados has kept the orthodox argument intact in writing an overview of the "unwinnable war" in Vietnam.[8]

The economic troubles in Vietnam after the Communist takeover coupled with Ronald Reagan's call in the 1980s to view America's intervention in Southeast Asia as "a noble cause" provided the spark for revisionist scholarship.[9] Beginning in the late 1970s, revisionists reexamined American military strategy, defended U.S. involvement in Southeast Asia, and blamed particular individuals and institutions for losing the war. In dissecting both the successful and unsuccessful elements of U.S. military strategy, revisionists began to shed more light on CAPs. In 1982, revisionist Col. Harry G. Summers argued that an official declaration of war from Congress and a strategy focused on annihilating the North Vietnamese Army would have brought victory for the United States in Vietnam. Summers highlights the tactical effectiveness of CAPs but ultimately contends that counterinsurgency units only strengthened the passive defensive strategy that handcuffed the U.S. military in Vietnam.[10] Contrary to Summers, Col. Andrew Krepinevich's *The Army and Vietnam* contends that the U.S. Army's institutional obsession with the war of attrition was the most significant contributor to America's ultimate failure in Vietnam. As part of Krepinevich's assault on the war of attrition, he gives a glowing appraisal of CAPs while blasting the U.S. Army for neglecting the program.[11] Although revisionist scholars provided the first multiparagraph descriptions of CAPs, the evolution of the program and the distinct social, cultural, and military dynamics of CAP villages remained overlooked in the general historiography of the war.

In 1989 Michael Peterson's *The Combined Action Platoons* became the first monograph in the historiography solely dedicated to the program. Peterson, a CAP veteran, offers a general overview of CAPs but admits that he was unable to research all necessary materials to create a thorough analysis. While he does venture into descriptions of CAP life from the perspective of the Marines and corpsmen in the villages, the majority of Peterson's book focuses on the administrative and management aspects of the program.[12] Nonetheless, *The Combined Action Platoons* proved to be an oft-cited reference for succeeding scholars drafting works on the military history of the Vietnam War.[13]

Publications based strictly on the Marines in Vietnam include descriptions of the program, yet the role of CAPs becomes lost within the framework of the entire Marine Corps effort.[14] Works specifically dedicated to CAPs come from veterans. Marine Vietnam veteran Al Hemingway compiled the oral histories of twenty-seven program veterans for his book, *Our War Was Different.* The veterans include NCOs and enlisted Marines and corpsmen who served in the villages as well as several colonels who managed the program.[15] Francis West's *The Village,* perhaps the most popular of the CAP veteran books, discusses social, political, cultural, and military obstacles CAP Marines faced in Binh Nghia village.[16] Likewise, Barry Goodson's memoir, *CAP MOT,* chronicles his daily routines in a CAP after the January 1968 Tet Offensive.[17] While these publications give readers unprecedented access into the inner workings of CAPs, they lack a broader interpretation of the program's relationship to other American and Vietnamese military and civilian groups.

Notes

Preface

1. Ricks, *Making the Corps,* 188.

Introduction

1. One should take note of this book's use of "Viet Cong." The Ngo Dinh Diem regime of the Republic of Vietnam (RVN) originally coined the term Viet Cong, which means "Vietnamese Communist," to identify the enemy political cadre, guerrillas, and insurgents who comprised the National Liberation Front (NLF) and its military arm, the People's Liberation Armed Forces (PLAF). Some scholars view the term as pejorative since various non-Communist South Vietnamese joined the NLF. The term was widely used by both RVN and U.S. civilian and military officials during the war. In writing from the perspective of the Marines and corpsmen in the program, who also used the term, I have chosen to use Viet Cong when referencing the NLF and PLAF. Moreover, the more conventional military force of the Democratic Republic of Vietnam (DRV) is known as both the North Vietnamese Army (NVA) and the People's Army of Vietnam (PAVN), and this work exclusively utilizes the former.

2. Trullinger, *Village at War.*

3. Milam, *Not a Gentlemen's War.* Milam served in Vietnam as a junior officer in the U.S. Army assigned to Mobile Advisory Team 38 in Pleiku.

4. Allnutt, "Marine Combined Action Capabilities," C-6.

5. Quoted in Moser, "To Keep a Village Free."

6. Critchfield, "The Marines Try a New Kind of Warfare."

7. Appy, *Working Class War,* 25.

8. Quoted in Ebert, *A Life in a Year,* 10.

9. Longley, *Grunts,* 14, 15.

10. The U.S. military separates the distinct jobs of soldiers, sailors, airmen, and Marines into various occupational fields. The U.S. Marine Corps has a four-digit numerical code that describes a Marine's occupational specialty. The first two num-

bers identify the general military field, such as infantry (03), artillery (08), or ordnance (05), and the last two digits describe the specific specialty. For example, (03) represents duty in the infantry, and the succeeding (11) designates that Marine infantryman specifically as a rifleman. Thus, a Marine with the designated MOS (0311) is a rifleman in the infantry, whereas (0331) is a machine gunner in the infantry.

11. In May 1965, the Marines complied with Gen. William C. Westmoreland's request to change the III Marine Expeditionary Force to III Marine Amphibious Force. Westmoreland feared that the Vietnamese would relate "expeditionary" to the French Expeditionary Force during the 1950s.

12. Clodfelter, *Vietnam in Military Statistics,* 252, 107.

13. FMFPAC monthly report, August 1969.

14. See Perry, "Marines in Afghanistan Take 'the Village' to Heart."

1. The Evolution of Combined Action Platoons

1. See Bickel, *MARS Learning.*

2. IIIMAF commanders in Vietnam such as Lewis Walt were aware of the manual but did not believe it possessed all the answers for winning the "small war" in Southeast Asia.

3. U.S. Marine Corps, *Small Wars Manual,* 1.

4. Isely and Crowl, *The U.S. Marines and Amphibious War,* 45–46. This work is the most comprehensive publication to date on the development and implementation of the Fleet Marine Force before and during World War II.

5. Quoted in Blanchard, "Pacification," 56.

6. Millett, *Semper Fidelis,* 548.

7. See Lee, *Utter's Battalion.* Lee, a retired lieutenant colonel in the Marine Corps, served in the Second Battalion, Seventh Marine regiment from August 1964 to June 1966. Lee witnessed firsthand the conventional amphibious assault training the Marine Corps had implemented prior to the beginning of U.S. combat involvement in Vietnam.

8. See Jablon, "General David M. Shoup."

9. Quoted in Millett, *Semper Fidelis,* 558.

10. U.S. Marine Corps, *Operations against Guerrilla Forces,* 110.

11. Roe et al., *A History of Marine Corps Roles and Missions,* 25.

12. See Krepinevich, *The Army and Vietnam.* Krepinevich coined the term "the Concept" to describe the army's dedication to conventional warfare in the years between Korea and Vietnam.

13. Westmoreland, *A Soldier Reports,* 156.

14. Puller is perhaps the most celebrated Marine in the history of the service. Even today, Marine recruits learn about the hard-nosed, aggressive attitude of Puller, known as the "Marine's Marine." See Davis, *Marine!* and Hoffman, *Chesty.*

15. Hoffman, *Chesty,* 117.

16. Walt, *Strange War, Strange Strategy,* 43.

17. Ibid., 63.

18. Ibid., 47.

19. FMFPAC monthly report, April 1966.

20. Walt, *Strange War, Strange Strategy,* 89.

21. Mullen, "Modifications to the IIIMAF Combined Action Program in the Republic of Vietnam," C-3.

22. Ibid., C-5.

23. Ibid., C-11.

24. Shulimson and Johnson, *U.S. Marines in Vietnam,* 138.

25. Raines, "An Analysis of the Command and Control Structure of the Combined Action Program," B-12.

26. See Corson, "Phong Bac Hamlet: Case Study in Pacification."

27. Each CAP had a numerical designation that referenced the group, company, and the specific platoon. For example, 4-3-2 refers to the second CAP of the third company of the fourth group.

28. Corson, *The Betrayal.*

29. FMFPAC monthly report, November 1967.

30. Quoted from Shulimson et al., *U.S. Marines in Vietnam,* 601.

31. Raines, "An Analysis of the Command and Control Structure of the Combined Action Program," B-19.

32. Marine Corps commandant, discussion with Ambassador Komer.

33. FMFPAC monthly report, February 1970. The reason for the decreased rating under the new HES ratings system was that under the previous IIIMAF rating scale, military security was stressed in the score. The HES rating system considered other political and economic factors.

34. Donnelly and Shore, *U.S. Marines in Vietnam.*

35. Bobrowsky, interview by Shulimson.

36. Quoted in Peterson, *The Combined Action Platoons,* 56.

37. Ibid., 61.

38. The glaring discrepancy in the casualty numbers stems from the often bloated enemy "body count" figures from the U.S. military. William Duiker argues that the enemy casualty rate likely hovered closer to thirty thousand, while Allan Millett cites sixty thousand as the probable figure. Duiker, *Sacred War,* 213; Millett, *Semper Fidelis,* 595.

39. Military History Institute of Vietnam, *Victory in Vietnam,* 238.

40. Cosmas and Murray, *U.S. Marines in Vietnam,* 149.

41. FMFPAC monthly report, March 1970.

42. Quoted in Peterson, *The Combined Action Platoons,* 77.

2. Combined Action Platoons, Green Berets, and Mobile Advisory Teams

1. The term *montagnards,* or "mountain people," originated with the French, who simply identified them according to the mountainous topography that characterized the Central Highlands where the ethnic minorities settled.

2. According to the U.S. Army Center of Military History's series on the army in Vietnam, the First Special Service Force in World War II is the antecedent to the SF in Vietnam. The First Special Service Force was created for use in snow-covered terrain. Its members received training in demolitions, rock climbing, amphibious assaults, and ski techniques. See Department of the Army, *U.S. Army Special Forces*.

3. See Donlon, *Beyond Nam Dong*.

4. For firsthand accounts of mobile guerrilla teams, see Yedinak, *Hard to Forget*; and Donahue, *Mobile Guerrilla Force*.

5. Quoted in Jackson, "The Vietnamese Revolution and the Montagnards," 328.

6. Quoted in Clarke, *Advice and Support*, 198.

7. Krepinevich, *The Army and Vietnam*, 72.

8. Memorandum from Westmoreland, "Basis for Discussion of Special Forces Employment."

9. Krepinevich, *The Army and Vietnam*, 70–74.

10. Moyar, *Phoenix and the Birds of Prey*, 37.

11. Weinraub, "8,000 Tribesmen Transplanted to Isolate the Vietcong."

12. Hickey, *Free in the Forest*, xix.

13. Ibid., xiv.

14. The French used the Central Highlands for various purposes, including as a place of exile for Vietnamese prisoners and a destination for montagnards forcibly moved to make room for plantations. See Pelley, *Postcolonial Vietnam*, 72–73.

15. Hickey, *Shattered World*, xiv.

16. Colby, *Lost Victory*, 283.

17. Jackson, "The Vietnamese Revolution and the Montagnards."

18. The name of the FULRO movement originated in French as Front unifié de lutte des races opprimées. Although the name in English (United Struggle Front for the Oppressed Races) does not match FULRO, Americans kept the original French-based acronym.

19. Raymond, "Elite Units Long a Source of Friction in the U.S. Army."

20. Hunt, *Pacification*, 106–7.

21. Westmoreland, interview by MacDonald.

22. Hunt, *Pacification*, 127.

23. FMFPAC monthly report, April 1969.

24. MACV, yearly summary.

25. See Donovan, *Once a Warrior King*, 64.

26. CORDS Headquarters, Territorial Division: IIIMAF-ICTZ, "Comparison of CAF, MAT, and CUPP Effectiveness."

27. Kreger, interview by Calkins.

28. McLeroy, interview by Maxner.

29. Donlon, *Beyond Nam Dong*, 117.

30. Kreger interview.

31. Stanton, *Green Berets at War*, 43.

32. Kreger interview.

33. Ibid.

34. Steve Stibbens, "Uphill Fight for the Medics."

35. Wilensky, *Military Medicine,* 29.

36. "U.S. Units Quit Viet Village."

37. Donlon, *Beyond Nam Dong,* 118.

38. Donovan, *Once a Warrior King,* 33, 331, 149.

39. Yarborough, "'Young Moderns' Are Impetus behind Army's Special Forces." According to Gen. William Yarborough, who served as the commandant of Fort Bragg's Special Warfare Center, the army rejected about one-third of the enlisted soldiers who volunteered for SF duty.

40. Training for medics lasted longer than any other specialty on an SF team.

41. Milam, interview by Verrone.

42. Harrison, memorandum to commanding general.

43. Program of Instruction: RF/PF Advisor Personnel.

44. Milam interview.

45. FMFPAC monthly report, September 1968.

46. Shulimson et al., *U.S. Marines in Vietnam,* 620.

47. FMFPAC monthly report, May 1968.

3. Becoming a Combined Action Platoon Marine

1. William Corson, in Hemingway, *Our War Was Different,* 50.

2. The historiography of the war features numerous authors who find Westmoreland's choice of search and destroy missions to fight a war of attrition ill suited to the political and military environment in Vietnam. See Krepinevich, *The Army and Vietnam;* Corson, *The Betrayal;* Sorley, *A Better War;* Sheehan, *A Bright Shining Lie;* Krulak, *First to Fight;* Prados, *Vietnam.*

3. Quoted in Personal Response Project, "Report of Survey Taken in IIIMAF."

4. Dower, *War without Mercy.* Dower's latest publication argues that "racial arrogance" and "cultural blindness" regarding Afghan fighters are part of the reason U.S. military and civilian leaders failed to take asymmetric threats seriously in Afghanistan. See W. Dower, *Cultures of War,* 131.

5. Linderman, *The World within War.*

6. Sledge, *With the Old Breed,* 34.

7. Roediger, *Towards the Abolition of Whiteness,* 117–20.

8. Quoted in Bradley, *Imagining Vietnam and America,* 48. Depending on the context, *Annam* can refer either to central Vietnam under French colonial rule or to the entire country. Modern writers, however, use the term to refer to the central part of Vietnam.

9. Ibid., 64.

10. O'Neil, interview by Maxner; Crawley, interview by Maxner; Wear, interview by Maxner.

11. Kindsvatter, *American Soldiers,* 194.

12. Crawley interview.

13. O'Neil interview.

14. Quoted in Shulimson et al., *U.S. Marines in Vietnam,* 616.

15. Esslinger, interview by Verrone.

16. Milam, *Not a Gentleman's War,* 100. These figures originally came from the "Tactical Initiative in Vietnam" study in the U.S. Army's Center of Military History Archives in Washington, DC.

17. Caputo, *A Rumor of War,* 113, 74. It should be noted that while Caputo is a Marine veteran of the Vietnam War, some believe his work to be a combination of fact and historical fiction.

18. Goller, interview by Southard.

19. Puller, *Fortunate Son,* 101–2, 117.

20. Hackworth, *Steel My Soldiers' Hearts,* 132.

21. Personal Response Project, "Report of Survey Taken in IIIMAF."

22. Trullinger, *Village at War,* 117.

23. Personal Response Project, "Report of Survey Taken in IIIMAF."

24. Caputo, *A Rumor of War,* 110, 134.

25. Ebert, *A Life in a Year,* 374.

26. Bobrowsky interview.

27. Goodson, *CAP MOT,* 18.

28. Kelly, letter to Lundquist.

29. Kelly, letter to fleet chaplain.

30. Chief of chaplains, letter to fleet chaplain.

31. Chief of chaplains, letter to chief of naval personnel.

32. McGonigal, letter to chief of chaplains.

33. The newsletter explained that Vietnamese Buddhists do not offer thanks because they believe a good deed earns a reward in the next life. Contrary to the American custom, the giver should thank the receiver for the privilege of earning a reward in the next life.

34. McGonigal, letter to chief of chaplains.

35. McGonigal, letter to director, Chaplain Corps Planning Group.

36. Personal Response Project, "Report of Survey Taken in IIIMAF."

37. McGonigal, letter to Craven.

38. Personal Response Project officer, letter to chief of chaplains.

39. Stevenson, letter to chief of chaplains.

40. Personal Response Project, "Report of American-Vietnamese Attitudes."

41. Schaedel, interview by Southard.

42. Personal Response Project, "Interim Report on Combined Action Units."

43. Peterson, *The Combined Action Platoons,* 42.

44. Second CAG command chronology, February 1969. In February 1969, Second CAG estimated that 150 of the 575 Marines and navy personnel in the group had not received the equivalent of a high school GED.

45. Shulimson et al., *U.S. Marines in Vietnam,* 617.

46. Sorley, *Vietnam Chronicles,* 316.

47. Ek, interview by Hunter.

48. The first CAP encompassed several villages, unlike future CAPs, which generally occupied only one.

49. Ek interview.

50. The interview was usually led by the director of the program and another CAG officer.

51. Tolnay, CAP questionnaire.

52. Tom Harvey, in Hemingway, *Our War Was Different,* 72.

53. Allnutt, "Marine Combined Action Capabilities," E-7.

54. Klyman, "The Combined Action Program."

55. Bobrowsky interview. Bobrowsky was also one of the few CAP Marines who did not attend CAP school.

56. The 80 CAPs existing at the close of November 1967 fell far short of plans for 114 by the end of the year.

57. Klyman, "The Combined Action Program," 13.

58. The command structure of the program featured CAPs at the lowest level, in the hamlets. Combined action companies (CACOs) at the district level controlled the multiple number of CAPs in their area of operations. Combined action groups (CAGs) administered the CACOs at the province level.

59. Tolnay questionnaire. The phrase "gone native" refers to American GIs who had distanced themselves from the "normal" soldier or Marine by assimilating into the Vietnamese culture.

60. Klyman, "The Combined Action Program," 12.

61. Goodson, *CAP MOT,* 15.

62. Bobrowsky interview.

63. Crawley interview.

64. Goller interview.

65. Not every Marine attended CAP school. During times of dire manpower need, the program would dispatch Marines into the villages without any CAP school training, but this was the exception, not the rule. An example of Marines completing only a portion of the education came with the 1968 Tet Offensive, which forced the school to send its students into the field after only one week of class.

66. Allnutt, "Marine Combined Action Capabilities," D-3.

67. One should note that these particular customs were not present in all the villages.

68. Goodson, *CAP MOT,* 18.

69. Corson, interview by Schioin and Greenman.

70. Jack Broz, in Hemingway, *Our War Was Different,* 128.

71. Corson, *The Betrayal,* 163.

72. Allnutt, "Marine Combined Action Capabilities," D-3, D-4.

73. Michael Cousino, in Hemingway, *Our War Was Different,* 97.

74. Critchfield, "The Marines Try a New Kind of Warfare."

75. Goodrich, interview by Maxner.

76. Assistant chief of staff, Combined Action Program, memorandum to chief, Territorial Forces Division.

77. First CAG command chronology, 1 March 1969–31 March 1969.

78. First CAG command chronology, 1 June 1969–30 June 1969.

79. Rocky Jay, in Hemingway, *Our War Was Different*, 115, 139.

80. In addition to the translation in Vietnamese, program leaders also realized the size and composition of combined action units paralleled that of a platoon rather than a company. The company level of the program changed from CAC to CACO.

81. Allnutt, "Marine Combined Action Capabilities," F-23, D-3.

82. Ibid., F-24.

83. Klyman, "The Combined Action Program," 8.

84. Goodson, *CAP MOT,* 17–18.

85. Allnutt, "Marine Combined Action Capabilities," C-3.

4. Life in a Combined Action Platoon

1. Ralph Smith, *Viet-Nam and the West*, 176.

2. Jamieson, *Understanding Vietnam*, 42.

3. Fitzgerald, *Fire in the Lake*, 8.

4. Nguyen Duy Hinh and Tran Dinh Tho, in Sorley, *The Vietnam War*, 720.

5. Fitzgerald, *Fire in the Lake*, 9, 15.

6. Jamieson, *Understanding Vietnam*, 28.

7. Ibid., 23–24.

8. Fitzgerald, *Fire in the Lake*, 10.

9. Pham, *Two Hamlets in Nam Bo*, 111.

10. Sorley, *The Vietnam War*, 723.

11. For a detailed analysis of Vietnamese rural life under French rule, see Long, *Before the Revolution.*

12. Lam, *Colonialism Experienced,* 49.

13. Jamieson, *Understanding Vietnam*, 91.

14. Pham, *Two Hamlets in Nam Bo,* 31, 32, 34.

15. Bradley, *Vietnam at War,* 79.

16. Ibid., 84.

17. The battle of Ap Bac in 1963, matching the ARVN against the VC in the first major confrontation between the two forces, is the iconic representation of the state of the South Vietnamese military under Diem shortly before his assassination in November. The ARVN units at Ap Bac ignored the advice from U.S. advisors, including the head advisor of the Seventh ARVN Division, John Paul Vann, who hovered over the battle in an observation aircraft, arguing that they did not take orders from Americans. The battle of Ap Bac was a disaster for the ARVN that foreshadowed the entire U.S. effort in Vietnam, as Vann's superiors ignored his attempts

to make them aware of the lackluster state of the South Vietnamese military and the need to change U.S. strategy. See Sheehan, *A Bright Shining Lie;* and Toczek, *The Battle of Ap Bac, Vietnam.*

18. Jamieson, *Understanding Vietnam,* 228.

19. Bradley, *Vietnam at War,* 87; Prados, *Vietnam,* 69.

20. Prados, *Vietnam,* 70.

21. Bradley, *Vietnam at War,* 97.

22. Trullinger, *Village at War,* 72.

23. Oka, "Saigon's Big Push."

24. JUSPAO, "Research Report."

25. Hop Brown, in Hemingway, *Our War Was Different,* 22.

26. Murphy, interview by Southard.

27. Kaupus, interview by Southard.

28. John A. Daube, in Hemingway, *Our War Was Different,* 120.

29. Longley, *Grunts,* 48.

30. Kaupus interview.

31. P. E. Dawson, in Klyman, "The Combined Action Program," 21.

32. Palm, "Tiger Papa Three."

33. Daube, in Hemingway, *Our War Was Different,* 120.

34. Kaupus interview.

35. Second CAG command chronology, June 1969. In June 1969, Second CAG reported that three hundred Vietnamese took advantage of the swimming lessons.

36. Critchfield, "The Marines Try a New Kind of Warfare." During the Tet holiday in 1967 in the CAP village of Hoa Phu, the Marines received multiple dinner invitations per day.

37. Bobrowsky interview.

38. Murphy interview.

39. Cousino, in Hemingway, *Our War Was Different,* 98.

40. Darby, "Combined Action Group Plays Hide and Seek with Charlie."

41. Schaedel interview.

42. Bing West, *The Village,* 181, 277, 217.

43. Schaedel interview.

44. Hickey, *Window on a War,* 207.

45. Independent Television News, *Roving Report,* n.d., in author's possession.

46. Department of the Navy, "Marines 'CARE.'"

47. Kaupus interview.

48. Stolfi, "US Marine Corps Civic Action Effort in Vietnam."

49. Thunhorst, interview by Southard.

50. Second CAG command chronology, January 1969.

51. Hinh and Tho, in Sorley, *The Vietnam War,* 720.

52. Thunhorst interview.

53. Chief, RF-PF Division, CAP graduation speech, 14 March 1970.

54. Allnutt, "Marine Combined Action Capabilities," 45.

55. Moser, "To Keep a Village Free"; Schaedel interview. Schaedel served in Echo-2 and was featured in the article, which was the cover story of this edition of *Life*.

56. Corson, *The Betrayal*, 204, 216.

57. Schaedel interview.

58. Telfer, Rogers, and Fleming, *U.S. Marines in Vietnam*, 184. In addition, the U.S. Marine Corps Reserves established a civic action fund for Vietnam. The Reserves fund collected cash-only donations used to purchase barber, carpenter, and midwifery tools and CARE packaged them into kits, shipped them, and distributed the finished products to Marine units, including CAPs.

59. Russell B. Longaway, e-mail correspondence with author, 7 December 2010.

60. Bing West, *The Village*, 186.

61. Ibid., 294, 260–61.

62. Ibid., 185.

63. Ibid., 240.

64. Warren V. Smith, in Hemingway, *Our War Was Different*, 139.

65. Palm, "Tiger Papa Three."

66. Bing West, *The Village*, 225.

67. Chuck Ratliff, in Hemingway, *Our War Was Different*, 28.

68. Allnutt, "Marine Combined Action Capabilities," 59.

69. For a history of MEDCAPs in Vietnam, see Wilensky, *Military Medicine to Win Hearts and Minds*.

70. Ibid., 53.

71. Herman, *Navy Medicine in Vietnam*, 59.

72. Second CAG command chronology, December 1968.

73. Allnutt, "Marine Combined Action Capabilities," 59.

74. U.S. Senate, Committee on the Judiciary, *Civilian Casualty and Refugee Problems in South Vietnam*.

75. Wayne Christiansen, in Hemingway, *Our War Was Different*, 130.

76. Department of the Army, Operations Report: Lessons Learned.

77. U.S. Marine Corps, Field Medical Service School, "General Medical Information (Vietnam)."

78. Pat Morris, interview by Southard.

79. Christiansen, in Hemingway, *Our War Was Different*, 133.

80. Fourth CAG command chronology, 1 July 1969. The specific time period for the corpsmen's absence is not given in the document.

81. Christiansen, in Hemingway, *Our War Was Different*, 132.

82. Allnutt, "Marine Combined Action Capabilities," 60.

83. Hickey, *Village in Vietnam*, 88.

84. Broz, in Hemingway, *Our War Was Different*, 126.

85. Englehorn, letters to his family.

86. Hickey, *Village in Vietnam*, 119–20.

87. Fourth CAG command chronology, February 1969.

88. Daube, in Hemingway, *Our War Was Different,* 122.
89. Second CAG command chronology, February 1969.
90. Metzger, "Combined Action Program Effectiveness."
91. FMFPAC monthly report, August 1969.
92. Second CAG command chronology, January 1969.
93. First CAG command chronology, 1 July 1968 to 31 October 1968.
94. Walt, *Strange War, Strange Strategy,* 94.
95. McHale, *Print and Power.*
96. Pike, *Viet Cong,* 155.
97. Tang, *A Viet Cong Memoir.*
98. A. W. Sundberg, in Hemingway, *Our War Was Different,* 111; Goller interview.
99. Trullinger, *Village at War,* 143.
100. Corson, "Marine Combined Action Program in Vietnam."
101. Bobrowsky interview.
102. VC propaganda leaflet.
103. Cousino, in Hemingway, *Our War Was Different,* 99.
104. Bing West, *The Village,* 178.
105. Fleet Marine Force Pacific, "The Marine Combined Action Program."
106. Schaedel interview.
107. Allnutt, "Marine Combined Action Capabilities," 42.
108. Department of the Navy, "Marines 'CARE.'"
109. Corson, "Marine Combined Action Program."
110. Kaupus interview.
111. Schaedel interview.
112. Mike Smith, interview by Southard; Molina, interview by Southard; Bennington, interview by Southard; Schaedel interview.
113. During the U.S. advisory years in Vietnam, Marines participated in "Father for a Day," similar to the CAPs adoption program. The "father" brought Vietnamese orphans to the local mess hall for Christmas dinners, gifts, and songs.
114. Averrill, interview by Southard.
115. Don Moser, cable to George P. Hunt, in *Life,* 25 August 1967, 3.
116. Daube, in Hemingway, *Our War Was Different,* 120.
117. Bennington interview.
118. Bing West, *The Village,* 63.
119. Warren V. Smith, in Hemingway, *Our War Was Different,* 137.
120. During the First Indochina War, the Viet Minh ordered villagers to kill their dogs so they wouldn't betray their movements to local French authorities. See Pham, *Two Hamlets in Nam Bo.*
121. Morton, interview by Southard; Goller interview.
122. Harvey, in Hemingway, *Our War Was Different,* 84.
123. E-mail correspondence with author, 7 March 2011.
124. Murphy interview.

125. Critchfield, "The Marines Try a New Kind of Warfare."
126. Trullinger, *Village at War,* 91, 105.
127. Joe Jennings, e-mail questionnaire given by author, 26 October 2010.
128. Schaedel interview.
129. Out of the 193,000 defectors, only 1,000 came from the NVA.
130. The name Kit Carson derived from Christopher "Kit" Carson, known as a rugged frontiersman in the American West during the nineteenth century.
131. Second CAG command chronology, December 1968.
132. Bing West, *The Village,* 195.
133. Bobrowsky interview.
134. Jim Shipp, e-mail correspondence with author, 26 October 2010.
135. Cousino, in Hemingway, *Our War Was Different,* 99.
136. Stolfi, "US Marine Corps Civic Action Effort in Vietnam."
137. Duffie, "I Keep It in My Heart and Wait for You."
138. Mike Smith interview.
139. Molina interview.
140. Harvey, in Hemingway, *Our War Was Different,* 84.
141. Longley, *Grunts,* 87.
142. Quoted in Kindsvatter, *American Soldiers,* 157, 159.

5. Popular Forces in Combined Action Platoons

1. III Marine Amphibious Force Headquarters, "Standard Operating Procedure for the Combined Action Program."
2. Critchfield, "The Marines Try a New Kind of Warfare."
3. FMFPAC monthly report, December 1968. Statistics from April to September 1968 reveal that even though PF in CAPs comprised only 17.5 percent of all PF in I Corps, they conducted 45.5 percent of all PF operations and inflicted 39 percent of the enemy losses among all those units in I Corps. FMFPAC monthly report, April 1970. During the first three months of 1970, with 114 CAPs (each with at least one PF platoon), and 825 non-CAP PF platoons in I Corps, the CAP PF accounted for nearly half the enemy killed and more than half of the total weapons captured. Colby, memorandum to deputy for CORDS corps advisors. CORDS reported that PF in I Corps had a kill ratio of 7.3:1. The next-highest kill ratio came from PF in IV Corps, at 3.7:1. That same month, PF in I Corps had 1,326 enemy contacts, compared with 1,601 enemy contacts for PF in IV Corps. Fisk, letter to deputy for CORDS. In 1970, CORDS recognized that PF in I Corps displayed superior leadership, lower desertion rates, and higher enemy kill ratios than the PF from any other corps tactical zone. Metzger, "Combined Action Program Effectiveness." More specifically within I Corps, the kill ratio for CAP PF (117 platoons) for the period January through September 1969 was more than double that of non-CAP PF (661 platoons).
4. Jennings questionnaire; William Silver, questionnaire given by author; Charles R. Lockwood, questionnaire given by author; Charlie Tsamardinos, ques-

tionnaire given by author; Morton interview; Goller interview; Schaedel interview; Murphy interview; Kaupus interview; Morris interview; Molina interview; Bennington interview; Mike Smith interview; Noa, interview by Southard.

5. The Simulmatics Corporation, "A Socio-Psychological Study of Regional/ Popular Forces in Vietnam," 5.

6. MACV, RF-PF handbook for advisors.

7. Westmoreland, *A Soldier Reports,* 263.

8. Moyar, *A Question of Command,* 155.

9. Wiest, *Vietnam's Forgotten Army,* 49. Wiest chronicles the lives of two ARVN soldiers, Pham Van Dinh and Tran Ngoc Hue, who traveled in two separate ideological directions towards the end of the war. Although both fought for South Vietnam, Hue continued to fight with ARVN while Dinh defected to the enemy. Much of the work is based on interviews Wiest conducted with Dinh and Hue.

10. Hunt, *Pacification,* 253.

11. Inter-agency Roles and Missions Study Group, report. GVN, or Government of Vietnam, is a reference to the South Vietnamese government.

12. Quoted in Nagl, *Learning to Eat Soup with a Knife,* 123.

13. Krepinevich, *The Army and Vietnam,* 25.

14. Wiest, *Vietnam's Forgotten Army,* 49, 74.

15. Moyar, *A Question of Command,* 155.

16. Simulmatics Corporation, "A Socio-Psychological Study of Regional/Popular Forces in Vietnam."

17. MACCORDS, I Corps field overview, May 1970.

18. MACCORDS, I Corps field overview, June 1970.

19. Bing West, *The Village,* 46, 247.

20. MACCORDS, province reports.

21. Firfer, letter to province senior advisor.

22. Wiest, *Vietnam's Forgotten Army,* 21.

23. Fisk, letter to deputy for CORDS.

24. Center for International Studies, "Achieving Pacification in Vietnam."

25. Wiest, *Vietnam's Forgotten Army,* 80.

26. Corson interview.

27. Wiest, *Vietnam's Forgotten Army,* 80.

28. Corson, *The Betrayal,* 87.

29. MACCORDS, "Territorial Security in Vietnam."

30. Wiest, *Vietnam's Forgotten Army,* 33.

31. Woodside, *Community and Revolution in Modern Vietnam,* 281.

32. Simulmatics Corporation, "A Socio-Psychological Study of Regional/Popular Forces in Vietnam," 5, 24. It is unknown whether any of the PF interviewed worked in a CAP. The interviewees came from various provinces in South Vietnam, with a larger focus on the Mekong Delta region.

33. Ibid., 11.

34. Central Intelligence Agency, "South Vietnam's Military Establishment."

35. Creighton Abrams to the Weekly Intelligence Estimate Update Council, 1 November 1969, in Sorley, *Vietnam Chronicles,* 291.

36. FMFPAC monthly report, November 1966. By the end of 1966, the program had fifty-eight CAPs.

37. FMFPAC monthly report, February 1966.

38. Bing West, *The Village,* 53.

39. MACV, RF-PF handbook for advisors.

40. Central Intelligence Agency, "South Vietnam's Military Establishment."

41. E. L. Lewis, letter to Cushman.

42. Akins, *Nam Au Go Go,* 114.

43. Allnutt, "Marine Combined Action Capabilities," 39.

44. Christansen, in Hemingway, *Our War Was Different,* 131.

45. Some PF departed the village for the entire day. See Fourth CAG command chronology, 1 April 1969–30 April 1969.

46. Allnutt, "Marine Combined Action Capabilities," 37.

47. Goodson, *CAP MOT,* 149.

48. Joseph Sullivan, conversation with Bing West, in West, *The Village,* 62. Sullivan's comments came after he had just taken command of the CAP at Binh Nghia.

49. Edward F. Danowitz, in Hemingway, *Our War Was Different,* 106. From October 1968 to April 1969, Danowitz served as the program's director.

50. Allnutt, "Marine Combined Action Capabilities," E-90.

51. Quoted in Moser, "To Keep a Village Free," 58.

52. Kaupus interview.

53. According to the incident report, the Marine who fired the shots that ultimately killed the PF also wounded a Marine with the same volley of fire.

54. Danowitz, letter to Cushman.

55. Bing West, *The Village,* 47.

56. Ratliff, in Hemingway, *Our War Was Different,* 31.

57. Mike Smith interview.

58. Bing West, *The Village,* 221.

59. Corson, *The Betrayal,* 193.

60. Bing West, *The Village,* 112.

61. Critchfield, "The Marines Try a New Kind of Warfare."

62. Bing West, *The Village,* 239.

63. Morton interview.

64. Duffie, "I Keep It in My Heart and Wait for You."

65. Among the countless conversations at CAP reunions and e-mail questionnaires this author conducted with CAP veterans, none expressed confidence in the PF after the Marines departed.

66. Francis T. McNamara, letter to Zais.

67. See Bing West, *The Village;* and FMFPAC monthly report, March 1970.

68. Bing West, *The Village,* 319.

69. Allnutt, "Marine Combined Action Capabilities," 63.

70. Wiest, *Vietnam's Forgotten Army,* 225.

71. CORDS, monthly province report: Thua Thien Province.

72. CORDS, Military Region I overview.

73. Wilbanks, "The Last 55 Days."

74. "Special Study of ARVN Collapse in Central Vietnam."

6. The Combined Action Program and U.S. Military Strategy in Vietnam

1. Coram, *Brute,* 290.

2. See Krepinevich, *The Army and Vietnam.*

3. Westmoreland, *A Soldier Reports,* 99. In his autobiography, Westmoreland revealed his dislike of the term "search and destroy." He believed the war's dissenters intentionally distorted the term, equating "search and destroy" with random, aimless American patrols in the countryside that ravaged Vietnamese villages.

4. Walt, *Strange War, Strange Strategy,* 96.

5. Critchfield, "The Marines Try a New Kind of Warfare."

6. Wagner, questionnaire given by Robert Klyman. From July 1967 to January 1968, Wagner served as the deputy director of the program. Corson, *The Betrayal.*

7. Tolnay to Robert Klyman, comments on senior honor's thesis.

8. Collier to Robert Klyman, comments on senior honor's thesis.

9. Westmoreland, *A Soldier Reports,* 175.

10. Sheehan, *A Bright Shining Lie,* 642.

11. Kinnard, *The War Managers,* 45.

12. Davidson, *Vietnam at War,* 411. For a revisionist account of Westmoreland's attachment to PROVN, see Birtle, "PROVN, Westmoreland, and the Historians."

13. Ricks, *Making the Corps,* 195.

14. Barlow, *Revolt of the Admirals,* 35.

15. Millett, *Semper Fidelis,* 533.

16. Ibid., 545.

17. McMaster, *Dereliction of Duty,* 82.

18. Ibid., 83.

19. Ibid., 144.

20. Soon after the landing of the Marines at Da Nang, the Marines changed "expeditionary" to "amphibious," or IIIMAF, because the former was thought to bring back bad memories for the Vietnamese of the previous French expeditionary forces.

21. One should note that during the buildup of U.S. Marines in Vietnam in 1965, U.S. Army general Maxwell Taylor, serving as a military consultant to President Johnson, threw his support behind the enclave strategy. Ultimately, however, Johnson and Westmoreland's tenacity in pursuing an offensive strategy quieted Taylor's push for the enclave strategy. See Halberstam, *The Best and the Brightest,* 574–84.

22. Quoted in Nagl, *Learning to Eat Soup with a Knife,* 157.

23. Coram, *Brute,* 294.

24. For a firsthand account of Harvest Moon, see Lee, *Utter's Battalion.*

25. Walt, *Strange War, Strange Strategy,* 127.

26. In addition to opposing the firing of Thi, demonstrators also protested the overall corruption prevalent in Saigon.

27. Walt, interview by Merick.

28. Military History Institute of Vietnam, *Victory in Vietnam,* 176.

29. Tuohy, "Marine's Leaders Disappoint."

30. Shulimson et al., *U.S. Marines in Vietnam,* 625.

31. Quoted from Krepinevich, *The Army and Vietnam,* 175.

32. Krulak, letter to Nitze.

33. Holbrooke, letter to Komer.

34. Gole, *General William E. DePuy,* 165, 157, 163.

35. Quoted from Krepinevich, *The Army and Vietnam,* 175.

36. Finney, "Gen. Krulak Urges Marines to Resist Detractors in Army."

37. Sheehan, *A Bright Shining Lie,* 295, 296.

38. Ibid., 305.

39. Coram, *Brute,* 272–74.

40. Krulak, *First to Fight,* 181.

41. Quoted from Sheehan, *A Bright Shining Lie,* 630.

42. Krulak, "A Strategic Appraisal," 13, 3, 17.

43. Quoted from Buzzanco, *Masters of War,* 251.

44. Krulak, *First to Fight,* 202.

45. Krulak, letter to McNamara, 9 May 1966; Krulak, letter to McNamara, 4 January 1967.

46. McNamara, *In Retrospect,* 243.

47. Walt, *Strange War, Strange Strategy,* 39, 43, 124.

48. Walt, address at the thirteenth annual reunion.

49. Walt, address at the First Marine Division Association Reunion.

50. Westmoreland, MACV yearly report.

51. Westmoreland, *A Soldier Reports,* 201, 176, 177, 202.

52. Ibid., 177.

53. Duiker, *Sacred War,* 207.

54. Westmoreland, *A Soldier Reports,* 179.

55. Corson, in Hemingway, *Our War Was Different,* 55.

56. Corson, *The Betrayal,* 176.

57. Ward Just, "It's 3-Front War in Viet I Corps Area."

58. Quote from Shulimson et al., *U.S. Marines in Vietnam,* 619.

59. Komer, "Pacification."

60. Quote from Shulimson et al., *U.S. Marines in Vietnam,* 625.

61. McNamara, letter to Zais, 14 March 1970.

62. Metzger, letter to Zais.

63. See Sorley, *A Better War.*

64. Ibid., 11.

65. Quoted in Sorley, *Thunderbolt*, 208–9.

66. See Nickerson, letter to Abrams. Nickerson, as commanding general of IIIMAF, wrote to Abrams about the state of the PF. In the memo, along with a brief description of the mission and purpose of CAPs, Nickerson frequently notes the success of the program in improving the effectiveness of the PF.

67. FMFPAC monthly report, May 1968.

68. Combined Action Force, fact sheet.

69. Murphy interview.

70. First CAG command chronology, November 1969; First CAG command chronology, December 1969; First CAG command chronology.

Conclusion

1. Williamson, "The U.S. Marine Corps Combined Action Program (CAP)."

2. Palm, "Tiger Papa Three."

3. Edward Palm, in Hemingway, *Our War Was Different*, 39.

4. Gordon, "Army, Marine Corps Unveil Counterinsurgency Field Manual."

5. Petraeus, Amos, and Nagl, *The U.S. Army / Marine Corps Counterinsurgency Field Manual*, 187.

6. In 2004, U.S. Marine general James Mattis asked CAP veterans from Vietnam to speak with battalion commanders about to embark to Iraq on how to successfully operate a combined unit. See Bing West, *The Strongest Tribe*, 29.

7. Ibid., 185.

8. See ibid.

9. Johnson, "A Reevaluation of the Combined Action Program as a Counterinsurgency Tool," 24. Johnson is currently an officer in the Marine Corps.

10. Goodale and Webre, "The Combined Action Platoon in Iraq."

11. Iscol, "CAP India."

12. Johnson, "Reevaluating the CAP."

13. Perry, "Marines in Afghanistan Take 'the Village' to Heart."

14. Moyar, "Getting Close to the Afghans."

Appendix

1. Davidson, *Vietnam at War;* Kolko, *Anatomy of a War.*

2. Karnow, *Vietnam;* Herring, *America's Longest War.*

3. Sorley, *A Better War.*

4. FitzGerald, *Fire in the Lake.* Fitzgerald references the Combined Action Program once to describe the numerous jobs held by Col. William Corson, the director of the program in 1967. Hunt, *Pacification.* In Hunt's analysis of pacification during the war, the acronym CAP emerges only when comparing the program to U.S. Army mobile advisory teams.

5. See Lewy, *America in Vietnam*, 116–17; Spector, *After Tet*, 189–96.

---

6. Halberstam, *The Making of a Quagmire*. Halberstam's 1964 account of the war was the first major American publication to doubt the possibility of American success in Vietnam.

7. Cooper, *The Lost Crusade*.

8. Prados, *Vietnam*.

9. Catton, "Refighting Vietnam in the History Books."

10. Summers, *On Strategy*, 175.

11. Krepinevich, *The Army and Vietnam*, 172–77.

12. Peterson, *The Combined Action Platoons*.

13. Spector, *After Tet*; Moyar, *Phoenix and the Birds of Prey*; Wiest, *Vietnam's Forgotten Army*.

14. Edward F. Murphy, *Semper Fi, Vietnam*; Millett, *Semper Fidelis*; Shulimson and Johnson, *U.S. Marines in Vietnam*; Shulimson, *U.S. Marines in Vietnam*; Telfer, Rogers, and Fleming, *U.S. Marines in Vietnam*; Shulimson et al., *U.S. Marines in Vietnam*; Charles R. Smith, *U.S. Marines in Vietnam*; Cosmas and Murray, *U.S. Marines in Vietnam*.

15. Hemingway, *Our War Was Different*.

16. Bing West, *The Village*.

17. Goodson, *CAP MOT*. Shortly after the 1968 Tet Offensive, Combined Action Program leaders shifted many CAPs from a static strategy, in which the Marines were stationed in one compound in the village, to a mobile strategy, which featured Marines constantly moving to different places within their area of operations to keep the enemy off balance.

Bibliography

Primary Sources

Akins, John. *Nam Au Go Go: Falling for the Vietnamese Goddess of War.* Port Jefferson, NY: Vineyard, 2005.

Allnutt, Bruce C. "Marine Combined Action Capabilities: The Vietnam Experience," December 1969. U.S. Marine Corps Archives and Special Collections, Quantico, VA.

Assistant chief of staff, Combined Action Program. Memorandum to chief, Territorial Forces Division, CORDS-III Marine Amphibious Force. "CAP Vietnamese Language School: Information concerning," 1 January 1970. Folder 15, box 7, Military Assistance Command Vietnam, RG 472, National Archives II, College Park, MD.

Averrill, Robert. Interview by John Southard, 12 November 2010. San Antonio, TX.

Baxter, Jay. Interview by John Southard, 9 November 2009. Washington, DC.

Bennington, Bill. Interview by John Southard, 8 November 2009. Washington, DC.

Biggar, Kaye. Interview by Laura M. Calkins, 26 January 2004. Item #OH0351, the Vietnam Archive, Texas Tech University, Lubbock, TX.

Bobrowsky, Igor. Personal interview by Jack Shulimson, 3 December 1982. Folder 7, box 1, Robert Klyman Collection, U.S. Marine Corps Archives and Special Collections, Quantico, VA.

Brady, Martin. Interview by Richard Burks Verrone, 25 June, 9 July, 20 August 2003. Item #OH0310, the Vietnam Archive, Texas Tech University, Lubbock, TX.

Caputo, Philip. *A Rumor of War.* New York: Owl, 1996.

Center for International Studies. "Achieving Pacification in Vietnam," 1968. Folder 1 C (3), box 60, NSF: Vietnam, Lyndon Baines Johnson Presidential Library, Austin, TX.

Central Intelligence Agency. "South Vietnam's Military Establishment: Prospects for Going It Alone," December 1968. Folder 1 C (4)-B, box 60, NSF: Vietnam, Lyndon Baines Johnson Presidential Library, Austin, TX.

Chief of chaplains. Letter to chief of naval personnel, 12 September 1966. Folder 2, box 17, Personal Response Project, U.S. Marine Corps Archives and Special Collections, Quantico, VA.

———. Letter to fleet chaplain, Pacific, 9 June 1966. Folder 2, box 17, Personal Response Project, U.S. Marine Corps Archives and Special Collections, Quantico, VA.

Chief, RF-PF Division: CORDS, III Marine Amphibious Force. CAP graduation speech, 14 March 1970. Folder 15, box 7, Military Assistance Command Vietnam, RG 472, National Archives II, College Park, MD.

Colby, William. *Lost Victory: A Firsthand Account of America's Sixteen-Year Involvement in Vietnam.* Chicago: Contemporary Books, 1989.

———. Memorandum to deputy for CORDS corps advisors, 17 May 1969. Folder 39, box 3, RG 472, National Archives II, College Park, MD.

Combined Action Force. Fact sheet, 31 March 1970. U.S. Marine Corps History Division, Vietnam War Documents Collection, the Vietnam Archive, Texas Tech University, Lubbock, TX.

CORDS, Territorial Division: IIIMAF-ICTZ. "Comparison of CAF, MAT, and CUPP Effectiveness in Improving Efficiency of RF/PF and Effect on Pacification," 20 January 1970. Folder 1605-04A, box 7, Military Assistance Command Vietnam, RG 472, National Archives II, College Park, MD.

———. Military Region I overview, April 1972. Folder 3, box 1, Glenn Helm Collection, the Vietnam Archive, Texas Tech University, Lubbock, TX.

———. Monthly province report: Thua Thien Province, March 1972. Folder 4, box 1, Glenn Helm Collection, the Vietnam Archive, Texas Tech University, Lubbock, TX.

Corson, William. Interview by Michelle Schioin and Ron Greenman, 12 October 1984. Robert Klyman Collection, U.S. Marine Corps Archives and Special Collections, Quantico, VA.

———. "Marine Combined Action Program in Vietnam," n.d. Folder 8, box 152, RG 127, National Archives II, College Park, MD.

———. "Phong Bac Hamlet: Case Study in Pacification." Folder 6, box 1, Robert Klyman Collection, U.S. Marine Corps Archives and Special Collections, Quantico, VA.

Crawley, David. Interview by Stephen Maxner, 27 February 2001. Item #OH0083, the Vietnam Archive, Texas Tech University, Lubbock, TX.

Critchfield, Richard. "The Marines Try a New Kind of Warfare." *Star,* 28 April 1967. Folder 1, box 9, Douglas Pike Collection: Unit 2—Military Operations, the Vietnam Archive, Texas Tech University, Lubbock, TX.

Danowitz, Ed. Letter to Robert Cushman, 31 January 1969. Folder 18, box 2, RG 472, National Archives II, College Park, MD.

Darby, Don. "Combined Action Group Plays Hide and Seek with Charlie." *Sea Ti-*

ger, 21 August 1970. Folder 2, box 1, Curtis Englehorn Collection, the Vietnam Archive, Texas Tech University, Lubbock, TX.

Department of the Army. Operations Report: Lessons Learned; Report 2–68, "Medical Lessons Learned," 16 April 1968. Folder 31, box 1, Robert M. Hall Collection, the Vietnam Archive, Texas Tech University, Lubbock, TX.

Department of the Navy. "Marines 'CARE': A Study of Humanitarianism," 27 October 1967. Folder 14, box 9, Douglas Pike Collection: Unit 2—Military Operations, the Vietnam Archive, Texas Tech University, Lubbock, TX.

Donahue, James C. *Mobile Guerrilla Force: With the Special Forces in War Zone D.* Annapolis: Naval Institute Press, 1996.

Donlon, Roger H. C. *Beyond Nam Dong.* Leavenworth, KS: N, 1998.

Donnelly, Maj. Thomas, and Capt. Moyers S. Shore II. *U.S. Marines in Vietnam,* part 6, "Ho Chi Minh's Gamble." Chapters 17–19, Historical Division, HQMC, 10 April 1970. Folder 5, box 1, Robert Klyman Collection, U.S. Marine Corps Archives and Special Collections, Quantico, VA.

Donovan, David. *Once a Warrior King: Memories of an Officer in Vietnam.* New York: Ballantine Books, 1985.

Donovan, Jim. Interview by Stephen Maxner, 10 August 2000. Item #OH0013, the Vietnam Archive, Texas Tech University, Lubbock, TX.

Duffie, Timothy A. "I Keep It in My Heart and Wait for You." www.capmarine .com.

Ek, Paul. Interview by D. J. Hunter, 24 January 1966. Item #USMC0046, the Vietnam Archive, Texas Tech University, Lubbock, TX.

Englehorn, Curtis. Letters to his family. Curtis Englehorn Collection, the Vietnam Archive, Texas Tech University, Lubbock, TX.

Esslinger Tom. Interview by Richard B. Verrone, 21 August 2003. Item #OH0333, the Vietnam Archive, Texas Tech University, Lubbock, TX.

Evins, Gary. Interview by John Southard, 6 November 2009. Washington, DC.

Finney, John F. "Gen. Krulak Urges Marines to Resist Detractors in Army." *New York Times,* 13 May 1968, 1. Folder 7, box 12, Douglas Pike Collection: Unit 2—Military Operations, the Vietnam Archive, Texas Tech University, Lubbock, TX.

Firfer, Alexander. Letter to province senior advisor, Quang Nam Province, 25 April 1969. Folder 11, box 4, RG 472, National Archives II, College Park, MD.

First CAG command chronologies, January 1968–September 1970. U.S. Marine Corps History Division, Vietnam War Documents Collection, the Vietnam Archive, Texas Tech University, Lubbock, TX.

Fisk, Eugene. Letter to deputy for CORDS, 16 May 1970. Folder 1, box 6, RG 472, National Archives II, College Park, MD.

Fleet Marine Force Pacific. "The Marine Combined Action Program, Vietnam, 1967." Folder 14, box 146, RG 127, National Archives II, College Park, MD.

Flynn, Robert. *A Personal War in Vietnam.* College Station: Texas A&M University Press, 1989.

FMFPAC. "The Marine Combined Action Program, Vietnam," n.d. Folder 14, box 146, RG 127, National Archives II, College Park, MD.

———. Monthly reports, March 1965–June 1971. U.S. Marine Corps History Division, Vietnam War Documents Collection, the Vietnam Archive, Texas Tech University, Lubbock, TX.

Fourth CAG command chronologies, November 1968–July 1970. U.S. Marine Corps History Division, Vietnam War Documents Collection, the Vietnam Archive, Texas Tech University, Lubbock, TX.

Goller, Jerry. Interview by John Southard, 11 November 2010. San Antonio, TX.

Goodale, Jason, and Jon Webre. "The Combined Action Platoon in Iraq." *Marine Corps Gazette,* April 2005, 35–42.

Goodrich, Anthony. Interview by Stephen Maxner, 11 April 2002. Item #OH0131, the Vietnam Archive, Texas Tech University, Lubbock, TX.

Goodson, Barry L. *CAP MOT: The Story of a Marine Special Forces Unit in Vietnam, 1968–1969.* Denton: University of North Texas Press, 1997.

Gordon, Michelle. "Army, Marine Corps Unveil Counterinsurgency Field Manual," 15 December 2006. www.army.mil/news.

Hackworth, David H. *Steel My Soldiers' Hearts: The Hopeless to Hardcore Transformation of 4th Battalion, 39th Infantry, United States Army, Vietnam.* New York: Simon & Schuster, 2002.

Harrison, Francis X. Memorandum to commanding general, U.S. Army, Vietnam, 27 August 1969. Folder 24, box 2, RG 472, National Archives II, College Park, MD.

Hemingway, Al. *Our War Was Different: Marine Combined Action Platoons in Vietnam.* Annapolis: Naval Institute Press, 1994.

Herman, Jan K. *Navy Medicine in Vietnam: Oral Histories from Dien Bien Phu to the Fall of Saigon.* Jefferson, NC: McFarland, 2009.

Holbrooke, Richard. Letter to Robert Komer, 19 August 1966. Folder Vietnam Provincial Reports, box 230, NSF: Vietnam, Lyndon Baines Johnson Presidential Library, Austin, TX.

Independent Television News. *Rover Report,* n.d.

Inter-agency Roles and Missions Study Group. Report, 24 August 1966. Folder 22, box 6, Douglas Pike Collection: Unit 1—Assessment and Strategy, the Vietnam Archive, Texas Tech University, Lubbock, TX.

Iscol, Zachary J. "CAP India." *Marine Corps Gazette,* January 2006, 55–61.

JUSPAO. "Research Report, Nationwide Hamlet Survey, Fourth Interim Summary Report—IV Corps," 13 December 1967. Folder 1 C (2), box 59, NSF: Vietnam, Lyndon Baines Johnson Presidential Library, Austin, TX.

Just, Ward. "It's 3-Front War in Viet I Corps Area as Marines Fight for Pacification." *Washington Post,* 13 April 1967. Folder 14, box 8, Douglas Pike Collection: Unit 2—Military Operations, the Vietnam Archive, Texas Tech University, Lubbock, TX.

Kaupus, Paul. Interview by John Southard, 6 November 2009. Washington, DC.

Kelly, James W. Letter to fleet chaplain, Pacific, 9 June 1966. Folder 2, box 17, Personal Response Project, U.S. Marine Corps Archives and Special Collections, Quantico, VA.

————. Letter to L. M. Lundquist, 16 August 1966. Folder 2, box 17, Personal Response Project, U.S. Marine Corps Archives and Special Collections, Quantico, VA.

Komer, Robert W. "Pacification: A Look Back." *Army,* June 1970, 20–29. Folder 3, box 16, Douglas Pike Collection: Unit 2—Military Operations, the Vietnam Archive, Texas Tech University, Lubbock, TX.

Kreger, Robert. Interview by Laura M. Calkins, 28 October 2005. Item #OH0448, the Vietnam Archive, Texas Tech University, Lubbock, TX.

Krulak, Victor. Letter to Paul H. Nitze, 17 July 1966. Folder 6, box 3, Robert Klyman Collection, U.S. Marine Corps Archives and Special Collections, Quantico, VA.

————. Letter to Robert S. McNamara, 9 May 1966. Folder 6, box 3, Robert Klyman Collection, U.S. Marine Corps Archives and Special Collections, Quantico, VA.

————. Letter to Robert S. McNamara, 4 January 1967. Folder 6, box 3, Robert Klyman Collection, U.S. Marine Corps Archives and Special Collections, Quantico, VA.

————. "A Strategic Appraisal," December 1965. Folder 6, box 3, Robert Klyman Collection, U.S. Marine Corps Archives and Special Collections, Quantico, VA.

Lee, Alex. *Utter's Battalion: 2/7 Marines in Vietnam, 1965–1966.* New York: Ballantine Books, 2000.

Lewis, E. L. Letter to Robert Cushman, 15 January 1969. Folder 18, box 2, RG 472, National Archives II, College Park, MD.

MACCORDS. I Corps field overview, May 1970. Folder CORDS Monthly Overviews, box 8, Military Assistance Command Vietnam, RG 472, National Archives II, College Park, MD.

————. I Corps Field overview, June 1970. Folder CORDS Monthly Overviews, box 8, Military Assistance Command Vietnam, RG 472, National Archives II, College Park, MD.

————. Province reports: Quang Nam, period ending 31 January 1968. U.S. Marine Corps History Division, Vietnam War Documents Collection, the Vietnam Archive, Texas Tech University, Lubbock, TX.

————. "Territorial Security in Vietnam," 1 January 1971. Folder 14, box 3, William Colby Collection, the Vietnam Archive, Texas Tech University, Lubbock, TX.

MACV. RF-PF handbook for advisors, 1969. Folder 1, box 3, Douglas Pike Collection: Unit 3—Civil Operations, Revolutionary Development Support, the Vietnam Archive, Texas Tech University, Lubbock, TX.

————. Yearly summary, part II, 1969. Folder 1, Bud Harton Collection, the Vietnam Archive, Texas Tech University, Lubbock, TX.

Marine Corps commandant, discussion with Ambassador Komer, 7 January 1968. Folder 6, box 3, Robert Klyman Collection, U.S. Marine Corps Archives and Special Collections, Quantico, VA.

McGonigal, R. A. Letter to chief of chaplains, 1 August 1966. Folder 5, box 17, Personal Response Project, U.S. Marine Corps Archives and Special Collections, Quantico, VA.

———. Letter to director, Chaplain Corps Planning Group, 11 October 1966. Folder 2, box 17, Personal Response Project, U.S. Marine Corps Archives and Special Collections, Quantico, VA.

———. Letter to John Craven, 1 December 1966. Folder 2, box 17, Personal Response Project, U.S. Marine Corps Archives and Special Collections, Quantico, VA.

McLeroy, James. Interview by Stephen Maxner, 3 October 2000. Item #OH0061, the Vietnam Archive, Texas Tech University, Lubbock, TX.

McNamara, Francis T. Letter to Melvin Zais, 14 March 1970. U.S. Marine Corps History Division, Vietnam War Documents Collection, the Vietnam Archive, Texas Tech University, Lubbock, TX.

McNamara, Robert S. In Retrospect: The Tragedy and Lessons of Vietnam. New York: Times Books, 1995.

Metzger, T. E. "Combined Action Program Effectiveness," 15 October 1969. Briefing to the primary deputy assistant secretary (systems analysis) and acting deputy assistant secretary (regional programs), Department of Defense. Folder 5, box 4, Military Assistance Command Vietnam, RG 472, National Archives II, College Park, MD.

———. Letter to Melvin Zais, 24 March 1970. U.S. Marine Corps History Division, Vietnam War Documents Collection, the Vietnam Archive, Texas Tech University, Lubbock, TX.

Milam, John R. Interview by Richard B. Verrone, 27, 28, 30 July, 23 August, 7 September 2005; 30, 31 January 2006. Item #OH0429, the Vietnam Archive, Texas Tech University, Lubbock, TX.

Molina, Jose. Interview by John Southard, 8 November 2009. Washington, DC.

Morris, Pat. Interview by John Southard, 6 November 2009. Washington, DC.

Morton, Tom. Interview by John Southard, 7 November 2009. Washington, DC.

Moser, Don. "To Keep a Village Free." Life Magazine, 25 August 1967, 24–29, 58–62.

Mullen, John J. "Modifications to the IIIMAF Combined Action Program in the Republic of Vietnam," 19 December 1968. Student staff study, Amphibious Warfare School, Quantico, VA. Folder 2, box 2, Robert Klyman Collection, U.S. Marine Corps Archives and Special Collections, Quantico, VA.

Murphy, Mike. Interview by John Southard, 6 November 2009. Washington, DC.

Nickerson, H. Letter to Creighton Abrams, 14 January 1970. Folder 15, box 7, Military Assistance Command Vietnam, RG 472, National Archives II, College Park, MD.

Noa, Michael. Interview by John Southard, 9 November 2009. Washington, DC.
O'Brien, Tim. *If I Die in a Combat Zone, Box Me Up and Ship Me Home.* New York: Broadway Books, 1975.
Oka, Takashi. "Saigon's Big Push." *Christian Science Monitor,* 19 March 1964. Folder 20, box 3, Douglas Pike Collection: Unit 2—Military Operations, the Vietnam Archive, Texas Tech University, Lubbock, TX.
O'Neil, Jack. Interview by Steve Maxner, 5, 10, 16 September 2002. Item #OH0147, the Vietnam Archive, Texas Tech University, Lubbock, TX.
Palm, Edward. "Tiger Papa Three." Draft copy for *Marine Corps Gazette.* U.S. Marine Corps Archives and Special Collections, Quantico, VA.
Perry, Tony. "Marines in Afghanistan Take 'the Village' to Heart." *Los Angeles Times,* 8 January 2010, A20.
Personal Response Project. "Interim Report on Combined Action Units, III Marine Amphibious Force, with Special Attention to Personal Relationships," 8 February 1967. Folder 9, box 17, Personal Response Project, U.S. Marine Corps Archives and Special Collections, Quantico, VA.
———. "Report of American-Vietnamese Attitudes at NCO Leadership School-Camp Hansen, Okinawa, October 1966 to June 1967." Folder 14, box 17, Personal Response Project, U.S. Marine Corps Archives and Special Collections, Quantico, VA.
———. "Report of Survey Taken in IIIMAF TAOR among USMC and USN Personnel to Determine Their Attitudes toward ARVN and PFs and the Indigenous Local People," September 1966. Folder 7, box 17, Personal Response Project, U.S. Marine Corps Archives and Special Collections, Quantico, VA.
Personal Response Project officer. Letter to chief of chaplains, 1 August 1967. Folder 6, box 17, Personal Response Project, U.S. Marine Corps Archives and Special Collections, Quantico, VA.
Petraeus, David H., James F. Amos, and John A. Nagl. *The U.S. Army / Marine Corps Counterinsurgency Field Manual.* Chicago: University of Chicago Press, 2006.
Pham, David Lan. *Two Hamlets in Nam Bo: Memoirs of Life in Vietnam through Japanese Occupation, the French and American Wars, and Communist Rule, 1940–1986.* Jefferson, NC: McFarland, 2000.
Program of Instruction: RF/PF Advisor Personnel, n.d. Folder 2, box 24, Military Assistance Command Vietnam, RG 472, National Archives II, College Park, MD.
Puller, Lewis, Jr. *Fortunate Son: An Autobiography.* New York: Grove, 1991.
Raymond, Jack. "Elite Units Long a Source of Friction in the U.S. Army." *New York Times,* 3 October 1964, 3. Folder 6, box 4, Douglas Pike Collection: Unit 2—Military Operations, the Vietnam Archive, Texas Tech University, Lubbock, TX.
Schaedel, Ron. Interview by John Southard, 11 November 2010. San Antonio, TX.
Second CAG command chronologies, October 1968–May 1971. U.S. Marine Corps

History Division, Vietnam War Documents Collection, the Vietnam Archive, Texas Tech University, Lubbock, TX.

Simulmatics Corporation. "A Socio-Psychological Study of Regional/Popular Forces in Vietnam," 1967. Box 239, NSF: Vietnam, Lyndon Baines Johnson Presidential Library, Austin, TX.

Siu, Lap. Interview by John Southard, 18 February 2011. Lubbock, TX.

Sledge, E. B. *With the Old Breed: At Peleliu and Okinawa.* New York: Ballantine Books, 1981.

Smith, Mike. Interview by John Southard, 9 November 2009. Washington, DC.

Sorley, Lewis. *Vietnam Chronicles: The Abrams Tapes, 1968–1972.* Lubbock: Texas Tech University Press, 2004.

"Special Study of ARVN Collapse in Central Vietnam, April 1975." Folder 13, box 25, Douglas Pike Collection: Unit 2—Military Operations, the Vietnam Archive, Texas Tech University, Lubbock, TX.

Stevenson, Neil. Letter to chief of chaplains, 2 April 1969. Folder 4, box 17, Personal Response Project, U.S. Marine Corps Archives and Special Collections, Quantico, VA.

Stibbens, Steve. "Uphill Fight for the Medics: Medicine Replaces Magic." *Stars and Stripes,* 12 February 1963, 7. Folder 5, box 3, Douglas Pike Collection: Unit 2—Military Operations, the Vietnam Archive, Texas Tech University, Lubbock, TX.

Stolfi, Russell H. "US Marine Corps Civic Action Effort in Vietnam, March 1965–March 1966," 1968. Folder 14, box 12, Douglas Pike Collection: Unit 11—Monographs, the Vietnam Archive, Texas Tech University, Lubbock, TX.

Tang, Truong Nhu. *A Viet Cong Memoir: An Inside Account of the Vietnam War and Its Aftermath.* New York: Vintage Books, 1985.

Taylor, Maxwell D. *Swords and Plowshares.* New York: Da Capo, 1972.

Third CAG command chronologies, December 1968–September 1970. U.S. Marine Corps History Division, Vietnam War Documents Collection, the Vietnam Archive, Texas Tech University, Lubbock, TX.

III Marine Amphibious Force Headquarters. "Standard Operating Procedure for the Combined Action Program," 6 December 1969. Folder 15, box 7, Military Assistance Command Vietnam, RG 472, National Archives II, College Park, MD.

Thunhorst, Richard. Interview by John Southard, 12 November 2010. San Antonio, TX.

Tolnay, J. J. CAP questionnaire, 6 November 1985. Folder 2, box 1, Robert Klyman Collection, U.S. Marine Corps Archives and Special Collections, Quantico, VA.

Tuohy, William. "Marine's Leaders Disappoint U.S. Command." *Washington Post,* 3 March 1968. Folder 13, box 11, Douglas Pike Collection: Unit 2—Military Operations, the Vietnam Archive, Texas Tech University, Lubbock, TX.

U.S. Congress. Senate. Committee on the Judiciary. *Civilian Casualty and Refugee Problems in South Vietnam.* 90th Cong., 2nd sess., 9 May 1968. Folder 6, box 31,

Douglas Pike Collection: Unit 11—Monographs, the Vietnam Archive, Texas Tech University, Lubbock, TX.

U.S. Marine Corps. *Operations against Guerrilla Forces,* 1962. Folder 4, box 18, Douglas Pike Collection: Unit 3—Insurgency Warfare, the Vietnam Archive, Texas Tech University, Lubbock, TX.

U.S. Marine Corps, Field Medical Service School. "General Medical Information (Vietnam)," n.d. Folder 6, box 2, Calvin Chapman Collection, the Vietnam Archive, Texas Tech University, Lubbock, TX.

VC propaganda leaflet, 20 April 1970. Folder 2, box 2, Robert Klyman Collection, U.S. Marine Corps Archives and Special Collections, Quantico, VA.

Wagner, David H. Questionnaire given by Robert Klyman, 25 August 1985. Folder 1, box 1, Robert Klyman Collection, U.S. Marine Corps Archives and Special Collections, Quantico, VA.

Walt, Lewis W. Address at the First Marine Division Association reunion, 29 July 1967. Folder 6, box 3, Robert Klyman Collection, U.S. Marine Corps Archives and Special Collections, Quantico, VA.

———. Address at the thirteenth annual reunion, Third Marine Division Association, 22 July 1967. Folder 6, box 3, Robert Klyman Collection, U.S. Marine Corps Archives and Special Collections, Quantico, VA.

———. Interview by Wendell S. Merick. *U.S. News & World Report,* 22 May 1967. Folder 4, box 9, Douglas Pike Collection: Unit 2—Military Operations, the Vietnam Archive, Texas Tech University, Lubbock, TX.

———. *Strange War, Strange Strategy: A General's Report on Vietnam.* New York: Award Books, 1970.

Wear, John. Interview by Stephen Maxner, 29 October 2002. Item #OH0232, the Vietnam Archive, Texas Tech University, Lubbock, TX.

Weinraub, Bernard. "8,000 Tribesmen Transplanted to Isolate the Vietcong." *New York Times,* 22 June 1967. Folder 7, box 9, Douglas Pike Collection: Unit 2—Military Operations, the Vietnam Archive, Texas Tech University, Lubbock, TX.

West, Bing. *The Village.* New York: Pocket Books, 1972.

Westmoreland, William. Interview by Charles B. MacDonald, 4–5 February 1973. Box 30, Papers of William Westmoreland, Lyndon Baines Johnson Presidential Library, Austin, TX.

———. MACV yearly report, 1966. Folder 4, box 6, Douglas Pike Collection: Unit 2—Military Operations, the Vietnam Archive, Texas Tech University, Lubbock, TX.

———. Memorandum, "Basis for Discussion of Special Forces Employment," 13 April 1964. Folder 3, box 1, Papers of William Westmoreland, Lyndon Baines Johnson Presidential Library, Austin, TX.

———. *A Soldier Reports.* Garden City, NY: Doubleday, 1976.

Yarborough, William. "'Young Moderns' Are Impetus behind Army's Special Forces." *Army Magazine,* 4 May 1965, 1–14. (Orig. pub. March 1962.) Folder 8, box 5, Douglas Pike Collection: Unit 2—Military Operations, the Vietnam Archive, Texas Tech University, Lubbock, TX.

Yedinak, Steven M. *Hard to Forget: An American with the Mobile Guerrilla Force in Vietnam.* New York: Ivy Books, 1998.

Secondary Sources

Appy, Christian G. *Working Class War: American Combat Soldiers and Vietnam.* Chapel Hill: University of North Carolina Press, 1993.

Bacevich, Andrew. *The New American Militarism: How Americans Are Seduced by War.* New York: Oxford University Press, 2005.

Barlow, Jeffrey G. *Revolt of the Admirals: The Fight for Naval Aviation, 1945–1950.* Washington, DC: Brassey's, 1998.

Beckett, Ian F. W., and John Pimlott, eds. *Armed Forces and Modern Counter-insurgency.* New York: St. Martin's, 1985.

Bergerud, Eric. *Red Thunder, Tropic Lightning: The World of a Combat Division in Vietnam.* New York: Penguin, 1993.

Bergsma, Herbert L. *U.S. Marines in Vietnam: Chaplains with the Marines in Vietnam, 1962–1971.* Washington, DC: History and Museums Division Headquarters, U.S. Marine Corps, 1985.

Bickel, Keith B. *Mars Learning: The Marine Corps' Development of Small Wars Doctrine, 1915–1940.* Boulder: Westview, 2001.

Birtle, Andrew J. "PROVN, Westmoreland, and the Historians: A Reappraisal." *Journal of Military History* 72, no. 4 (2008): 1213–47.

Blanchard, Don H. "Pacification: Marine Corps Style." Master's thesis, Naval War College, 1968.

Blaufarb, Douglas S., and George T. Tanham. *Who Will Win? A Key to the Puzzle of Revolutionary War.* New York: Crane Russak, 1989.

Bradley, Mark Philip. *Imagining Vietnam and America: The Making of Postcolonial Vietnam, 1919–1950.* Chapel Hill: University of North Carolina Press, 2000.

———. *Vietnam at War.* Oxford: Oxford University Press, 2009.

Brigham, Robert. *ARVN: Life and Death in the South Vietnamese Army.* Lawrence: University Press of Kansas, 2006.

Buzzanco, Robert. *Masters of War: Military Dissent and Politics in the Vietnam Era.* Cambridge: Cambridge University Press, 1996.

Cable, Larry E. *Conflict of Myths: The Development of American Counterinsurgency Doctrine and the Vietnam War.* New York: New York University Press, 1986.

Carland, John M. "Winning the Vietnam War: Westmoreland's Approach in Two Documents." *Journal of Military History* 68, no. 2 (2004): 553–74.

Cassidy, Robert M. *Counterinsurgency and the Global War on Terror: Military Culture and Irregular War.* Westport, CT: Praeger Security International, 2006.

Catton, Phillip E. "Refighting Vietnam in the History Books: The Historiography of the War." *OAH Magazine of History* 18, no. 5 (2004): 7–11.

Cincinnatus. *Self-Destruction: The Disintegration and Decay of the United States Army during the Vietnam Era.* New York: Norton, 1981.

Clarke, Jeffrey J. *Advice and Support: The Final Years, 1965–1973.* The U.S. Army in Vietnam. Washington, DC: Center of Military History, 1988.

Clodfelter, Michael. *Vietnam in Military Statistics: A History of the Indochina Wars, 1772–1991.* Jefferson, NC: McFarland, 1995.

Collier, Tom, to Robert Klyman. Comments on senior honor's thesis, 28 January 1986. Folder 14, box 1, Robert Klyman Collection, U.S. Marine Corps Archives and Special Collections, Quantico, VA.

Cooper, Chester. *The Lost Crusade.* New York: Dodd, Mead, 1970.

Coram, Robert. *Brute: The Lie of Victor Krulak, U.S. Marine.* New York: Little, Brown, 2010.

Corson, William R. *The Betrayal.* New York: Norton, 1968.

Cosmas, Graham A., and Terrence P. Murray. *U.S. Marines in Vietnam: Vietnamization and Redeployment, 1970–1971.* Washington, DC: History and Museums Division Headquarters, U.S. Marine Corps, 1986.

Davidson, Phillip B. *Vietnam at War: The History, 1946–1975.* New York: Oxford University Press, 1988.

Davies, Bruce. *The Battle at Ngok Tavak: Allied Valor and Defeat in Vietnam.* Lubbock: Texas Tech University Press, 2008.

Davis, Burke. *Marine! The Life of Chesty Puller.* New York: Bantam Books, 1962.

Department of the Army. *U.S. Army Special Forces, 1961–1971.* Vietnam Studies. Washington, DC: U.S. Army Center of Military History, 1989.

DePauw, John W., and George A. Luz, eds. *Winning the Peace: The Strategic Implications of Military Civic Action.* New York: Praeger, 1992.

Dower, John W. *Cultures of War: Pearl Harbor / Hirsohima / 9–11 / Iraq.* New York: Norton, 2010.

———. *War without Mercy: Race and Power in the Pacific War.* New York: Pantheon Books, 1986.

Doyle, Edward, and Stephen Weiss. *A Collision of Cultures.* The Vietnam Experience. Boston: Boston Publishing, 1984.

Duiker, William J. *Sacred War: Nationalism and Revolution in a Divided Vietnam.* Boston: McGraw-Hill, 1995.

Ebert, James R. *A Life in a Year: The American Infantryman in Vietnam.* New York: Ballantine Books, 1993.

Ferguson, Ernest B. *Westmoreland: The Inevitable General.* Boston: Little, Brown, 1968.

Fisher, Christopher T. "The Illusion of Progress: CORDS and the Crisis of Modernization in South Vietnam, 1965–1968." *Pacific Historical Review* 75, no. 1 (2006): 25–51.

Fitzgerald, Frances. *Fire in the Lake: The Vietnamese and the Americans in Vietnam.* Boston: Little, Brown, 1972.

Gardner, Lloyd C., and Marilyn B. Young, eds. *Iraq and the Lessons of Vietnam; or, How Not to Learn from the Past.* New York: New Press, 2007.

Gole, Henry G. *General William E. DePuy: Preparing the Army for Modern War.* Lexington: University Press of Kentucky, 2008.

Halberstam, David. *The Best and the Brightest.* New York: Ballantine Books, 1969.

———. *The Making of a Quagmire.* New York: Random House, 1964.

Hennessy, Michael Alphonsus. "Divided They Fell: America's Response to Revolutionary War in I Corps, Republic of Vietnam, 1965–1971." Master's thesis, University of New Brunswick, 1989.

Herring, George C. "American Strategy in Vietnam: The Postwar Debate." *Military Affairs* 46, no. 2 (1982): 57–63.

———. *America's Longest War: The United States and Vietnam, 1950–1975.* 2nd ed. New York: Knopf, 1986.

Hickey, Gerald Cannon. *Free in the Forest: Ethnohistory of the Vietnamese Central Highlands, 1954–1976.* New Haven: Yale University Press, 1982.

———. *Shattered World: Adaptation and Survival among Vietnam's Highland Peoples during the Vietnam War.* Philadelphia: University of Pennsylvania Press, 1993.

———. *Village in Vietnam.* New Haven: Yale University Press, 1964.

———. *Window on a War: An Anthropologist in the Vietnam Conflict.* Lubbock: Texas Tech University Press, 2002.

Hoffman, Jon T. *Chesty: The Story of Lieutenant General Lewis B. Puller, USMC.* New York: Random House, 2002.

Hughes, Richard. "A Much Maligned Strategy? Search and Destroy Operations in Quang Ngai Province and the Mekong Delta, 1967–1969." PhD diss., University of Salford, 2001.

Hunt, Richard A. *Pacification: The American Struggle for Vietnam's Hearts and Minds.* Boulder: Westview, 1995.

Isely, Jeter A., and Philip A. Crowl. *The U.S. Marines and Amphibious War: Its Theory, and Its Practice in the Pacific.* Princeton: Princeton University Press, 1951.

Jablon, Howard. "General David M. Shoup, U.S.M.C.: Warrior and War Protestor." *Journal of Military History* 60, no. 3 (1996): 513–38.

Jackson, Larry R. "The Vietnamese Revolution and the Montagnards." *Asian Survey,* May 1969. Folder 9, box 13, Douglas Pike Collection: Unit 2—Military Operations, the Vietnam Archive, Texas Tech University, Lubbock, TX.

Jamieson, Neil L. *Understanding Vietnam.* Berkeley: University of California Press, 1993.

Johnson, Katie Ann. "Reevaluating the CAP." *Marine Corps Gazette,* June 2009, 24–27.

———. "A Reevaluation of the Combined Action Program as a Counterinsurgency Tool." Senior honor's thesis, Georgetown University, 2008. www.capmarine.com/cap/data.htm.

Karnow, Stanley. *Vietnam: A History.* New York: Penguin Books, 1984.

Kelly, Francis J. *U.S. Army Special Forces, 1961–1971.* Washington, DC: GPO, 1973.

Kindsvatter, Peter S. *American Soldiers: Ground Combat in the World Wars, Korea, and Vietnam.* Lawrence: University Press of Kansas, 2003.

Kinnard, Douglas. *The War Managers.* Annapolis: Naval Institute Press, 1977.

Klyman, Robert A. "The Combined Action Program: An Alternative Not Taken." Senior honor's thesis, University of Michigan, 1986. www.capmarine.com.

Kolko, Gabriel. *Anatomy of a War: Vietnam, the United States, and the Modern Historical Experience.* New York: New Press, 1985.

Krepinevich, Andrew F. *The Army and Vietnam.* Baltimore: Johns Hopkins University Press, 1986.

Krulak, Victor H. *First to Fight: An Inside View of the U.S. Marine Corps.* Annapolis: Naval Institute Press, 1984.

Lam, Truong Buu. *Colonialism Experienced: Vietnamese Writings on Colonialism, 1900–1931.* Ann Arbor: University of Michigan Press, 2000.

Latham, Michael E. "Redirecting the Revolution? The USA and the Failure of Nation-Building in South Vietnam." In "From Nation-Building to State Building," special issue, *Third World Quarterly* 27, no. 1 (2006): 27–41.

Lewy, Guenter. *America in Vietnam.* Oxford: Oxford University Press, 1978.

Linderman, Gerald F. *The World within War: America's Combat Experience in World War II.* Cambridge, MA: Harvard University Press, 1997.

Long, Ngo Vinh. *Before the Revolution: The Vietnamese Peasants under the French.* New York: Columbia University Press, 1973.

Longley, Kyle. *Grunts: The American Combat Soldier in Vietnam.* Armonk, NY: M. E. Sharpe, 2008.

Malarney, Shaun Kingsley. *Culture, Ritual, and Revolution in Vietnam.* Honolulu: University of Hawai'i Press, 2002.

Malkasian, Carter. "Toward a Better Understanding of Attrition: The Korean and Vietnam Wars." *Journal of Military History* 68, no. 3 (2004): 911–42.

Marston, Daniel, and Carter Malkasian, eds. *Counterinsurgency in Modern Warfare.* Oxford: Osprey, 2008.

McHale, Shawn Frederick. *Print and Power: Confucianism, Communism, and Buddhism in the Making of Modern Vietnam.* Honolulu: University of Hawai'i Press, 2004.

McMaster, H. R. *Dereliction of Duty: Lyndon Johnson, Robert McNamara, the Joint Chiefs of Staff, and the Lies That Led to Vietnam.* New York: Harper Perennial, 1997.

McWilliams, Timothy S., and Kurtis P. Wheeler, eds. *Al-Anbar Awakening: American Perspectives.* Vol. 1. Quantico, VA: Marine Corps University, 2009.

Milam, Ron. *Not a Gentleman's War: An Inside View of Junior Officers in the Vietnam War.* Chapel Hill: University of North Carolina Press, 2009.

Military History Institute of Vietnam. *Victory in Vietnam: The Official History of the People's Army of Vietnam.* Translated by Merle L. Pribbenow. Lawrence: University Press of Kansas, 2002.

Millett, Allan R. *Semper Fidelis: The History of the United States Marine Corps.* New York: Free Press, 1980.

Mockaitis, Thomas R. *Iraq and the Challenge of Counterinsurgency.* Westport, CT: Praeger Security International, 2008.

Montgomery, Gary W., and Timothy McWilliams, eds. *Al-Anbar Awakening: Iraqi Perspectives.* Vol. 2. Quantico, VA: Marine Corps University, 2009.

Moyar, Mark. "Getting Close to the Afghans." *Washington Times,* 1 March 2010.

———. *Phoenix and the Birds of Prey: Counterinsurgency and Counterterrorism in Vietnam.* Lincoln: University of Nebraska Press, 1997.

———. *A Question of Command: Counterinsurgency from the Civil War to Iraq.* New Haven: Yale University Press, 2009.

———. *Triumph Forsaken: The Vietnam War, 1954–1965.* Cambridge: Cambridge University Press, 2006.

Murphy, Edward F. *Semper Fi, Vietnam: From Da Nang to the DMZ, Marine Corps Campaigns, 1965–1975.* New York: Ballantine Books, 1997.

Nagl, John A. *Learning to Eat Soup with a Knife: Counterinsurgency Lessons from Malaya and Vietnam.* Chicago: University of Chicago Press, 2002.

Palmer, Bruce. *The 25-Year War: America's Military Role in Vietnam.* Lexington: University Press of Kentucky, 1984.

Palmer, Dave Richard. *Summons of the Trumpet: U.S.-Vietnam in Perspective.* San Rafael, CA: Presidio, 1978.

Pelley, Patricia M. *Postcolonial Vietnam: New Histories of the National Past.* Durham: Duke University Press, 2002.

Peterson, Michael E. *The Combined Action Platoons: The U.S. Marines' Other War in Vietnam.* New York: Praeger, 1989.

Pike, Douglas. *Viet Cong: The Organization and Techniques of the National Liberation Front of South Vietnam.* Cambridge: MIT Press, 1966.

———. *War, Peace, and the Viet Cong.* Cambridge: MIT Press, 1969.

Popking, Samuel L. "Pacification: Politics and the Village." In "Vietnam: Politics, Land Reform and the Development in the Countryside," special issue, *Asian Survey* 10, no. 8 (1970): 662–71.

Prados, John. *Vietnam: The History of an Unwinnable War, 1945–1975.* Lawrence: University Press of Kansas, 2009.

Race, Jeffrey. *War Comes to Long An: Revolutionary Conflict in a Vietnamese Province.* Berkeley: University of California Press, 1972.

Raines, R. C. "An Analysis of the Command and Control Structure of the Combined Action Program." Individual Research Project, U.S. Marine Corps Command and Staff College, 1969. Folder 4, box 2, Robert Klyman Collection, U.S. Marine Corps Archives and Special Collections, Quantico, VA.

Rambo, A. Terry. *Searching for Vietnam: Selected Writings on Vietnamese Culture and Society.* Kyoto: Kyoto University Press, 2005.

Ricks, Thomas E. *Making the Corps.* New York: Touchstone, 1997.

Roe, Thomas G., Ernest H. Guisti, John H. Johnstone, and Benis M. Frank. *A History of Marine Corps Roles and Missions: 1775–1962.* Washington, DC: Historical Branch, U.S. Marine Corps, 1962.

Roediger, David. *Towards the Abolition of Whiteness: Essays on Race, Politics and Working Class History.* New York: Verso, 1994.

Rottman, Gordon L. *The US Army in the Vietnam War, 1965–73.* New York: Osprey, 2008.

Salemink, Oscar. *The Ethnography of Vietnam's Central Highlanders: A Historical Contextualization, 1850–1900.* Honolulu: University of Hawai'i Press, 2003.

Sarkesian, Sam C. *Unconventional Conflicts in a New Security Era: Lessons from Malaya and Vietnam.* Westport, CT: Greenwood, 1993.

Schwab, Orrin. *A Clash of Cultures: Civil-Military Relations during the Vietnam War.* Westport, CT: Praeger Security International, 2006.

Scoville, Thomas W. *Reorganizing for Pacification Support.* Washington, DC: Center of Military History, 1982.

Sheehan, Neil. *A Bright Shining Lie: John Paul Vann and America in Vietnam.* New York: Vintage Books, 1988.

Shulimson, Jack. *U.S. Marines in Vietnam: An Expanding War, 1966.* Washington, DC: History and Museums Division Headquarters, U.S. Marine Corps, 1982.

Shulimson, Jack, Leonard A. Blasiol, Charles Smith, and Capt. David A. Dawson. *U.S. Marines in Vietnam: The Defining Year, 1968.* Washington, DC: History and Museums Division Headquarters, U.S. Marine Corps, 1997.

Shulimson, Jack, and Charles M. Johnson. *U.S. Marines in Vietnam: The Landing and the Buildup, 1965.* Washington, DC: History and Museums Division Headquarters, U.S. Marine Corps, 1978.

Sigler, David Burns. *Vietnam Battle Chronology: U.S. Army and Marine Corps Combat Operations, 1965–1973.* Jefferson, NC: McFarland, 1992.

Smith, Charles R. *U.S. Marines in Vietnam: High Mobility and Standown, 1969.* Washington, DC: History and Museums Division Headquarters, U.S. Marine Corps, 1988.

Smith, Ralph. *Viet-Nam and the West.* London: Heinemann, 1968.

Solis, Gary D. *Son Thang: An American War Crime.* New York: Bantam Books, 1997.

Sorley, Lewis. *A Better War: The Unexamined Victories and Final Tragedy of America's Last Years in Vietnam.* San Diego: Harcourt, 1999.

———. *Honorable Warrior: General Harold K. Johnson and the Ethics of Command.* Lawrence: University Press of Kansas, 1998.

———. *Thunderbolt: General Creighton Abrams and the Army of His Times.* Bloomington: Indiana University Press, 1992.

———, ed. *The Vietnam War: An Assessment by South Vietnamese Generals.* Lubbock: Texas Tech University Press, 2010.

Spector, Ronald. *Advice and Support: The Early Years of the U.S. Army in Vietnam, 1941–1960.* New York: Free Press, 1985.

———. *After Tet: The Bloodiest Year in Vietnam.* New York: Vintage Books, 1993.

Stanton, Shelby L. *Green Berets at War: U.S. Army Special Forces in Southeast Asia, 1956–1975.* New York: Ivy Books, 1985.

―――. *The Rise and Fall of an American Army: U.S. Ground Forces in Vietnam, 1965–1973*. New York: Ballantine Books, 1985.

Suhre, Christopher Gordon, "The United States Marine Corps' Counterinsurgency Effort in the Vietnam War." Master's thesis, Texas Tech University, 1998.

Summers, Harry G. *On Strategy: A Critical Analysis of the Vietnam War*. New York: Presidio, 1982.

Telfer, Gary L., Lane Rogers, and V. Keith Fleming. *U.S. Marines in Vietnam: Fighting the North Vietnamese, 1967*. Washington, DC: History and Museums Division Headquarters, U.S. Marine Corps, 1984.

Tierney, John J. *Chasing Ghosts: Unconventional Warfare in American History*. Washington, DC: Potomac Books, 2006.

Toczek, David. *The Battle of Ap Bac, Vietnam: They Did Everything but Learn from It*. Annapolis: Naval Institute Press, 2001.

Tolnay, John, to Robert Klyman. Comments on senior honor's thesis, 6 January 1986. Folder 10, box 1, Robert Klyman Collection, U.S. Marine Corps Archives and Special Collections, Quantico, VA.

Trullinger, James W. *Village at War: An Account of Conflict in Vietnam*. Stanford: Stanford University Press, 1994.

U.S. Marine Corps. *Small Wars Manual*. Washington, DC: Skyhorse, 2009.

"U.S. Units Quit Viet Village After 7-Year Pacification." *Sun,* 3 February 1969. Folder 12, box 13, Douglas Pike Collection: Unit 2—Military Operations, the Vietnam Archive, Texas Tech University, Lubbock, TX.

Vadas, Robert E. *The Vietnam War*. Cultures in Conflict. Westport, CT: Greenwood, 2002.

Weigley, Russell F. *The American War of War: A History of United States Military Strategy and Policy*. Bloomington: Indiana University Press, 1973.

West, Bing. *The Strongest Tribe: War, Politics, and the Endgame in Iraq*. New York: Random House, 2009.

West, Francis J. *Small Unit Action in Vietnam, Summer 1966*. New York: Arno, 1967.

Whitlow, Robert H. *U.S. Marines in Vietnam: The Advisory and Combat Assistance Era, 1954–1964*. Washington, DC: History and Museums Division Headquarters, U.S. Marine Corps, 1977.

Wiest, Andrew. *Vietnam's Forgotten Army: Heroism and Betrayal in the ARVN*. New York: New York University Press, 2008.

Wilbanks, James H. *The Battle of An Loc*. Bloomington: Indiana University Press, 2005.

―――. "The Last 55 Days, A Paper from the 3rd Triennial Vietnam Symposium." Third Triennial Symposium of the Vietnam Center, Texas Tech University, Lubbock, 15–17 April 1999. Folder 2, box 1, James Wilbanks Collection, the Vietnam Archive, Texas Tech University, Lubbock, TX.

―――. *The Tet Offensive: A Concise History*. New York: Columbia University Press, 2007.

Wilensky, Robert J. *Military Medicine to Win Hearts and Minds: Aid to Civilians in the Vietnam War.* Modern Southeast Asia Series. Lubbock: Texas Tech University Press, 2004.

Williamson, Curtis L. "The U.S. Marine Corps Combined Action Program (CAP): A Proposed Alternative Strategy for the Vietnam War." Student staff paper, U.S. Marine Corps Command and Staff College, 2002.

Woodside, Alexander B. *Community and Revolution in Modern Vietnam.* Boston: Houghton Mifflin, 1976.

Index

www.ingramcontent.com/pod-product-compliance
Lightning Source LLC
Chambersburg PA
CBHW030304100426
42812CB00002B/557